Japan – Culture of Wood

Japan – Culture of Wood

Buildings, Objects, Techniques

Christoph Henrichsen

Birkhäuser – Publishers for Architecture
Basel · Berlin · Boston

Japanese terms and names are transcribed using the Hepburn system. Long vowels are marked with a ^. Japanese terms appear in lower case here – with the exception of names. The most important tools and techniques appear in the text in transcribed form. Japanese does not divide words, the characters are placed without word spacings; for ease of reading and better understanding hyphens have been used in longer transcriptions.

Proper names are given in the usual Japanese order, with the forename following the family name.

The illustration on the dust jacket shows one arch of the Kintaikyô bridge in Iwakuni. Okamura Tatsuya's photograph was made available by the building site office.

The chapter block titles were drawn with a brush by Wada Junko.

This book is also available in a German language edition (ISBN 3-7643-7021-1)

A CIP catalogue record for this book is available from the Library of Congress, Washington D.C., USA

Bibliographic information published by Die Deutsche Bibliothek
Die Deutsche Bibliothek lists this publication in the Deutsche Nationalbibliografie; detailed bibliographic data is available in the Internet at <http://dnb.ddb.de>.

© 2004 Birkhäuser – Publishers for Architecture, P.O.Box 133, CH-4010 Basel, Switzerland
Part of Springer Science + Business Media

Printed on acid-free paper produced from chlorine-free pulp. TCF ∞

Printed in Germany

ISBN 3-7643-7022-X
www.birkhauser.ch

9 8 7 6 5 4 3 2 1

Layout and cover design:
Atelier Fischer, Berlin

Translation:
Michael Robinson, London

Lithography:
Bildpunkt, Berlin

Printing:
Medialis, Berlin

Binding:
Kunst- und Verlagsbuchbinderei Leipzig

Inhalt

Foreword

I have been interested in wood and the hand tools used for working with wood since I was a child. My grandmother was almost obsessed with wood, and bought me the first tools of my own. I visited an old cabinet-maker who did not have a single machine in his workshop almost every day. Woodworking tools were my first contact with Japan as well: I was introduced to them by a Japanese master violin-maker who had settled in Germany. At that time I started to think I would like to visit that distant country, to have a look round and collect a stock of tools. I did an apprenticeship as a cabinet-maker and spent two years in English furniture workshops before finally travelling to Japan on the Trans-Siberian Express in 1987. I cycled round almost all of the main island of Honshû and visited a good 40 workshops, mainly belonging to toolsmiths, cabinet-makers and carpenters. I formed mixed impressions. The craftsmen's entirely internalized movements seemed like a dignified theatrical performance, and I was fascinated by the precision of their work, but at the same time it was somehow depressing that most of them were old men, and the end of their tradition seemed to be in sight.

I did not immediately commit myself to a Japanese workshop as I had originally intended, but went back to Europe to study Japanese, among other things. I continued to visit wood craftsmen sporadically during a two-and-a-half year study visit to Tokyo and three years working on a Japanese temple-building site. Aware of the fact that many trades and crafts were threatened with extinction, I decided in 2001 to embark upon a systematic documentation of Japanese woodworking professions. My aim was to present techniques that were largely unknown in the West, and the intellectual and spiritual approach to the craft. I intended to trace the processes by which products emerged and describe the qualities of those products, not as instructions to be copied, but as an inspiration to everyone who works in wood and is interested in design and creativity.

After classifying the uses of wood as a material I looked for suitable workshops to demonstrate them. These were to represent a particular trade as clearly as possible, and make products of higher than average quality, chiefly by hand. Where possible, I chose workshops that had been paid little or no attention hitherto. This was because I quickly found out that craftsmen who had become well-known through the media were taking on new habits not typical of a craftsman. Usually craftsmen work so swiftly and almost automatically that it is difficult to capture all the working steps the first time round. If necessary you may have to ask them to stop, or to repeat a step. Craftsmen who have turned performers are different, they stop of their own accord while working and may even ask why you are not taking a photograph. Craftsmen who performed like this are not included in the book as a matter of principle.

I was glad to be able to engage Roland Bauer as a photographer who had often taken photographs of craftsmen at work and handled himself very sensitively in this context. He accompanied me on two of a total of six research periods extending over several weeks. He came to my attention through a documentation of rural crafts in the Hohenlohe region that he had published jointly with Frieder Stöckle. His photographs are not just of high documentary quality, they also convey the atmosphere of the workshops, impressively and accurately.

Buying a product was usually the best way into the research. The craftsman sensed that they had been acknowledged and understood, and opened up; sometimes I commissioned a product specially, a sliding door, for example. So in the course of the project I built up a collection of Japanese wooden products. A small selection of these can be seen in the colour illustrations to the introduction. Where possible, every stage of making a product was recorded. With relatively simple objects this is sometimes possible in a single day, but not in the case of complex constructions and in the building trade. Here the workshops or building sites were visited several times if possible, to cover the most important steps.

Unknown workshops that felt valued by this sudden attention were particularly co-operative. The fact that a foreign woodworker was taking an interest in them probably made them more communicative and less inhibited than they would have been if fel-low-countrymen had been doing the research. A very few workshops refused to take part, as they feared that publicizing their techniques might lead to plagiarism. Work with cherry bark, for instance, is particularly threatened by cheap imitations. The relevant workshops are concentrated in the small northern Japanese town of Kakunodate; it was not possible to conduct the research there for that reason, so a workshop away from this centre had to be found.

Birkhäuser Publishers turned out to be an ideal partner, being equally interested in the technical, creative and also the atmospheric qualities of the project. Editor Andreas Müller accompanied the emergence of the book, from research to production, with consistent interest, watchfulness and critical goodwill. Of about 60 workshops investigated, 30 were selected for reasons of space and time. These are all workshops in which production by hand is still largely practised. The description focuses mainly on how the products are made; this is preceded by information on the history of the product and the profession. Over and above this specialist information, my intention was to convey the surroundings and atmosphere of the workshops, the latter above all through the photographs.

My intention is to record the diversity of Japanese woodworking craft and stimulate interest in knowledge and skills that are under threat. This outside attention will make an impact in Japan and make it more likely that the crafts will survive. The techniques may seem antiquated and out of date at first, but the understanding of materials that shows behind them and the aesthetic qualities of the products have something to say to us today. The wooden objects are exemplary pieces of creativity: they are striking in form and sensitively detailed, functional and also durable because they age well. They are made as individual pieces or in small series and thus form a contrast to the mass-production of today. Above all they heighten our appreciation of the aesthetic qualities and expressive possibilities of wood as a material, which has come to the fore again over recent decades as a renewable resource, even though the dominant industrial processes usually treat it like an artificial material.

Introduction

Japan is seen world-wide as the epitome of traditional timber building. The country stands out by using wood for a variety of purposes in all spheres of life. Japan shares this central significance of wood with many other cultures, but nowhere else has woodworking been divided up into so many professions and specialist approaches. A long process of continuing specialization has led to unique knowledge about working and designing with wood. This development resulted in technically perfect products that use the material with full respect of its characteristic features. The products seem modern because their design often evokes minimalism. What is more, the preference for particularly attractive grain shows a unique sensitivity to the beauty of the material.

Kasuga-gongen-engi-emaki scroll dating from 1309, showing a building site.

(Omni-)presence of tree and wood

The availability of the material is a prerequisite for the wide variety of uses of wood in Japan. The Japanese archipelago was originally almost completely wooded. During the last 2000 years man has made more room for himself by clearing timber, being forced up into the mountains from the plains as a result, but even today almost 70 % of the country is wooded. It is only in the last 100 years that the original diversity of tree species has been reduced considerably. Deciduous woodland in particular has tended to be replaced by monocultures of Japanese cedar and cypresses.

Large areas of Europe associate timber building with poverty. Stone, ideally dressed stone or rendered brick, has been preferred for public and prestigious building projects. And even when wood was used, the building often followed the forms of a stone structure, thus simulating affluence. Timber building was increasingly cut back in Europe from the Middle Ages to the 20th century, and finally restricted to rural architecture, outbuildings and concealed structures, whereas in Japan even the most prestigious buildings like temples and palaces were built in wood from the earliest days until the late 19th century. Stone and tiles were known, but their use was restricted to a limited number of areas. Stone might appear in foundations, to clad platforms or as a floor covering. Tiles were used for hard roofs, in isolated cases for flooring and – though only as a second use – to build enormous enclosing walls. Last but not least the dominance of wood in architecture is shown by the fact that the few stone structures like small bridges or mausoleums often imitate wooden buildings.

Concentrating on wood as a building material was promoted for at least three reasons. Suitable building timber is available in such lavish quantities that there was no need to look for alternatives. Though granite, sandstone and tuff all occur as viable building stone they could be quarried in only a few parts of the country and would therefore have needed transporting over large distances, which would have been particularly difficult in such a mountainous country. The frequent earthquakes also worked in favour of timber construction. The very pliable structures used in wooden Japanese buildings, with few reinforcing diagonal bracings, distorted in minor earthquakes, but rarely collapsed, and could also be readily pushed back into shape.

One special feature of the Japanese culture of wood is that round timbers are frequently used, right down to the present day. Round timbers are ideal for roof beams, which are subject to high tensile stress, and for cantilever beams in eaves. The precision with which round timbers are used in Japanese building is unique in the world. Thanks to markings based on a centre line, the joints are just as true and engage as positively as members with a rectangular cross-section.

As wood was used almost exclusively even for the most prestigious building projects, planning and working techniques became highly refined, especially in the woodworking professions. Something else that promoted Japanese carpenters' almost proverbial precision and perfect surface treatment is that the structure is not usually clad or painted. It is true that when Buddhism was adopted in the mid 6th century, continental architecture, as typical of China, was taken over for temples and palaces: coloured architecture with structural timbers in red, white panels and end grain picked out in yellow. But although this kind of finish was initially applied to temples and public building projects, coloured structures made up a very small proportion of timber construction overall. Architecture using untreated, visible timber as a material – *shiraki* in Japanese (white wood) – makes correspondingly greater demands for precise work on the surface of the material.

Another domain in which wood was used was to make containers and utensils for preparing food, storage, transport or to keep valuables in. Here the variety of techniques used for constructing the containers is surprising. They can be carved from a single piece of wood (*kuri-mono*), fashioned by bending thinly split wood (*mage-mono*), turned on a lathe (*hiki-mono*) or assembled from boards (*sashi-mono*).

Many other uses of wood are almost forgotten today. Until the late 19th century, wood was by far the most important source of energy in Japan. Today, even in the country, wood is used for heating only in exceptional cases. In the 7th and 8th centuries thinly split wooden tablets (*mokkan*) were used to convey information and as a substitute for expensive paper.[1] Wood was also used in civil and hydraulic engineering, building dams and locks, for example.[2] Means of transport like boats, carts and sedan-chairs were made of wood into the late 19th century, and so was most household, craft and agricultural equipment.[3, 4]

Use is made of all parts of the tree. The trunk is used for round or cut timbers, and with certain species other parts are used as well: the rootstock (for panels, because of its spectacular grain, or for turning), the bark (for roof coverings, to protect outer walls, cherry bark to make thin-walled containers or to "sew" chip boxes), the twigs (deutzia twigs for wooden pegs, shoots from the eucalyptus bush for toothpicks) and sometimes even leaves (for polishing, Kashi and cherry leaves for wrapping Japanese sweetmeats). The use of wood is reflected in the written language, which was taken over from China in the 5th century. Most characters, which were originally pictograms, have two components, one indicates the meaning and the other pronunciation. Frequently recurring elements are known as radicals, 214 being identified in all. Radical no. 75, for "tree" appears in a particularly large number of characters, and it is almost always on the left. In this family of signs, the link with trees or wood as a material is usually direct. Semantically, characters containing the tree radical can be divided into four groups: tree species, parts of the tree, timber construction components and objects made of wood.

木

The character for tree and wood. As a radical, it indicates the meaning of many complex signs.

枢 柿 柱 框 栓 桁 梁 棟 欄
机 杓 杯 枡 桶 椀 棋 槌 棺
棚 楯 槍 櫂 樽 橋 機 櫃 櫛

Wooden building components
toboso – *door*
kokera – *wooden shingle*
hashira – *post/support*
kamachi – *frame, lintel*
sen – *wooden peg*
keta – *purlin*
hari – *beam*
mune – *ridge*
obashima – *handrail*

Lines below:

Wooden objects
tsukue – *desk*
shaku – *ladle*
hai – *cup*
masu – *gauge*
bachi – *drumstick*
oke – *tub*
wan – *bowl*
gi – *Japanese chess set*
tsuchi – *mallet*
hitsugi – *coffin*
tana – *shelf*
tate – *sign*
kai – *paddle*
taru – *barrel*
hashi – *bridge*
hata – *loom*
hitsu – *chest/case*
kushi – *comb*

Tree of the gods – cypress in the Kasuga Shrine precinct in Nara.

As they are of Chinese origin, these characters first of all provide evidence of the use of wood in ancient China. But among the characters containing the tree radical in particular there are strikingly many that first emerged in Japan, so-called *koku-ji*. Most of them are names for particular kinds of trees, but they also include coinages for woodworking professions (*moku* – craftsman; *soma* – woodcutter) or wooden parts or objects (*waku* – frame; *masu* – bushel; *kase* – bobbin).

Finally, trees also have a religious significance. In Japan's independent religious tradition, Shintoism, nature is inhabited by countless deities (*kami*). Most of them live in objects that are kept in shrines like mirrors or swords, but frequently parts of nature also count as their embodiments or symbols, and alongside mountains and rocks these also include trees, called *shim-boku* (divine tree). These trees are identified in the precinct of the shrine by a straw rope (*shime-nawa*), which is tied around the trunk and has zig-zag strips of paper hanging down from it (*shide*).

Shôji from the Suzuki workshop, Kyoto. The sliding doors are covered with paper on one side and are very light. Often a solid panel is set in grooves in the lower section.

How woodworking techniques and professions developed

Wood was first worked with sharpened stone blades. They were fastened parallel with or at right angles to wooden handles, and were the first adzes (*yoko-ono*) and axes (*tate-ono*). Then, after rice cultivation was adopted from the 3rd century BC onwards, the dominant tools were iron axes, chisels and slightly curved blades attached to long handles (*yari-ganna*, literally lance plane; planing knife). The wooden objects from this period that have been found in excavations, in Toro near Shizuoka, for example, all show signs of work with iron blades.[5] Thanks to the new tools it was now possible to create fingerlap joints, halved joints as well as mortise and tenon, all of which were used when building the earliest rice stores. The number of woodworking tools increased considerably in the 4th and 5th century: the burial mounds called *kofun* yielded gouges, hammers, drills and small saws, as well as axes, adzes and chisels.[6]

Buddhism was adopted in the mid 6th century and the central government was further strengthened, which led to many major temple and palace building projects. The Hôryû-ji buildings near Nara, which date from the late 7th century, now the oldest surviving timber buildings in the world, provide evidence of the high level to which timber construction had developed, and traces of work also indicate the tools and working methods used. The trunks were marked out with a snap line after felling, split with wedges and finally worked with adzes. If a particularly smooth surface was needed, or a precise one in the case of joints, the wood was smoothed with a planing knife. There are numerous traces of sawing, though only when cutting across the grain.

Wooden objects from the Shôsôin.

Objects belonging to Tenno Shômu, who died in 752, were stored in the Shôsôin repository built near the Tôdai-ji temple in Nara in the mid 8th century. A large number of these have survived, including many wooden objects and some tools. The shapes and above all the materials, like luxury tropical woods and ivory, suggest that some of the objects came from the continent. Even so they give us a clear picture of the state of woodcraft at the time.[7]

From the mid 13th century onwards the history of the foundation of shrines and temples was often also presented on narrative scrolls. Some show building sites, for example the Kasuga-gongen-engi-emaki scroll of 1309, which is famous for the richness of its detail. It shows batter boards for setting foundation stones and the most important steps in processing wood, from marking the round timbers with square and snap line, splitting the boards to smoothing with planing knives. Most activities were carried out sitting down, moving the tool towards the body, both approaches being characteristic features of Japanese woodworking. Saws also appear, but only small ones with lancet blades, as used for cutting to length or wood joints. Large frame saws for cutting trunks,

Small three-drawer
chest from the
Fukuhara workshop,
Kyoto. The corners of
the dovetailed carcass
are rounded, the
front edge was
finished with Urushi
varnish.

Small four-drawer
chest from the
Fukuhara workshop,
Kyoto. The carcass
sides taper towards
the top. The front is
closed by a lid with
a knob.

Logging with a large frame saw wielded by two workers, c. 1500 (Suntory Museum of Art, Tokyo).

Cutting a jacked-up beam into boards, woodcut by Hiroshige, c. 1830.

Carpenters, woodcut c. 1860.

manipulated by two craftsmen, do not appear until the 15th century. Until this time, wood varieties like cypress and Japanese cedar, which were easy to split, had been predominantly used in timber building. Cutting with saws made it possible to use different varieties that were difficult to split, like Keyaki and pine. The independent profession of sawyer, called *ôga-biki* after the large frame saw, also came into being.

The tool that is now most readily associated with Japanese woodworking, the plane with a fixed iron, appears as the last group of hand tools. They do not make their widespread appearance until the 16th century. At first they were pushed with two hands, like their

Cooper's goods from the Tanaka workshop, Ôdate. In the back row, two containers for pickling vegetables, in the front row a lidded container for serving boiled rice, a low scoop for sushi rice and a large bath ladle.

Rice wine barrel from the Tanaka workshop, Ôdate. The conical carcass with Japanese cedar staves is held together with hoops of split and twisted strip bamboo.

15

Chinese models, but the handle behind the iron was soon abandoned, and the plane was pulled.

Rapid growth of cities and political stability in the Edo period (1603–1868) led to relative affluence and greater demand, both leading to increased specialization and the emergence of numerous new woodworking professions. The seven-volume illustrated encyclopaedia of professions, Jinrinkin-môzui, dating from 1690, shows comb-makers, chip box-makers, mask carvers, musical instrument builders, and even craftsmen who specialized in toothpicks.[8] The number of tools also grew considerably along with increasing specialization.

The range of woodworking professions reached a high point at the turn of the 19th to the 20th century, after the 1868 Meiji restoration had embarked upon a course leading to modernization of the country following Western models. The adoption of Western techniques and products created a whole new series of specialisms. The 1300 pages of "Uses of wood for craft purposes" (Ki no kôgeiteiki riyô), published by the forestry department of the Ministry of Agriculture and Trade in 1912, which lists most wooden products and their centres, provide a useful survey. It also describes new crafts like constructing casting moulds or making letter cases, bentwood furniture and clock casings.[9]

Mechanical woodworking started in Japan in 1863, when a steam-powered sawmill was opened in Hakunodate in northern Japan. In 1872, a workshop was set up in Sapporo with woodworking machines imported from Britain. But wood was still cut by hand in some regions until the post-war years.

Wood trades depicted in an illustrated encyclopaedia of professions in 1690: box-makers, coopers, chip box-makers (top row from left); comb-makers, zither-makers, toothpicks (bottom row).

Sanbô from the Hyôta workshop, Ise. Round tray for food offerings at the Great Shrine at Ise. The high edge of the tray, and the top and bottom rings on the foot are made of thin strips of bent cypress wood, and sewn together with strips of cherry bark.

Lidded box from the Hyôta workshop, Ise. The body of this box, used for storing ritual objects, is rounded at the corners and sewn at the overlap. Several wedge-shaped incisions are made at the back to realize the narrow radius at the corners.

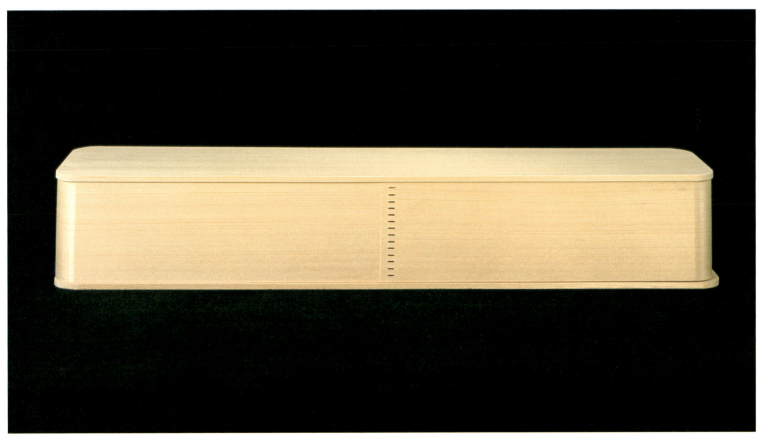

17

After the Second World War almost the entire traditional craft collapsed. Until then, following one's father into a profession had been taken for granted, now it became rare. Industrialization and the emergence of new materials like aluminium and plastics drove out a lot of wooden products and professions almost overnight. For example, the wide availability of rice cookers led to a collapse in demand for containers for cooked rice called *ohitsu*, which coopers assembled from staves. Intense specialization and a high level of identification with their products had also made Japan's craftsmen specialize mentally. They achieved great mastery and a very strong sense of self-confidence when making "their" product. But at the same time it was almost impossible for them to devise new products and apply their skills to them. As industry offered adequate employment in the post-war years thanks to rapid economic growth, many craftsmen abandoned their professions. The first electrical woodworking machines were made in Japan in 1957, and in the building trade in particular, most rough preparation work was soon handled mechanically.

Characteristics of Japanese wood products

Forms and dimensions
Japanese wood products show a marked preference for the simple basic stereometric forms cube, cylinder and cone. This is particularly clear in the case of containers like chip boxes, for example, bark boxes, vessels made of staves, packing cases, small pieces of furniture and chests of drawers. But even such varied products as wooden sandals, planes and game boards share this cubic form. Beyond these basic forms, the field is also dominated by simple and very striking outlines, in the case of lacquer bowls, for example, or Kokeshi dolls. Differently from Europe, Japanese architecture and applied art rarely use complex mouldings to upgrade or decorate (one of the rare uses of such contours made up of plinth, flute, curve and ogee moulding is for certain altars in temples). Plainness here is not to be confused with a lack of ideas, sensitivity or elegance. The products appeal in their precision, and show refinement and fine detailing when examined closely. From supporting members in architecture via the bars in a sliding door and on to household equipment the edges are almost always precisely bevelled, in other words chamfered off at an angle of 45°, or in rare cases 30°. This offers protection against mechanical damage, reduces cross-sections visually, making things look

Asymmetrical division of planes, example of an 18th century chest of drawers.

Cherry-bark tea cad-
dies from the
Ogasawara work-
shop, Tashiro.
The bark is smoothed
and polished for the
right-hand box. The
left-hand box shows
the relief-like, natural
cherry bark.

19

less heavy, and doubles the number of lines. Quite often slight, scarcely perceptible curves give the product a feeling of lightness and elegance, for example in the rising lines of temple eaves or the slightly bulging lids of high-quality packing cases. Such concave or convex outlines are evidence of the craftsmen's creative sensitivity and also of their mastery, as the lines run without a break, always full of tension, like a bent osier twig.

Also striking is the asymmetrical layout in both two and three dimensions. It is evident on every scale, from architecture to everyday objects. The earliest temple complexes were conceived symmetrically, like their continental models, arranged around a central axis running north-south, but already in the late 7th century in the Hôryû-ji temple the strict symmetry is broken by the five-storey pagoda opposite the main hall, which covers a much smaller area, but is higher. Asymmetrical arrangements occur particularly frequently in house building. This applies to the distribution of buildings within the complex as a whole but also to the ground plan and to the interior, for picture niches and writing alcoves, for example. Also, chests of drawers, which have been made in various variants and sizes since the 18th century, are often divided up irregularly at the front.

Lattice-work is particularly favoured. It is found especially frequently in moving parts of the interior, hence in the numerous variants of sliding doors and windows. The lattice can – as in the paper-covered *shôji* – be relatively fine, with a broad mesh, but often the bars are placed so close together that they occupy over half the area. In comparison with European wood products it is striking that extremely thin cross-sections are frequently used in Japan, so that the products are often very light. Reducing cross-sections in this way and thus also the amount of material used can be seen just as clearly in interior design, as in the case of door panels only a few millimetres thick, sliding door frames, ceiling panels or in most containers. Minimized cross-sections and lightweight construction are also always an indication of highly developed craft techniques.

Choice of material

Japan's woodworkers show intimate familiarity with their material, they know its qualities and are extremely sensitive to its aesthetic potential. Even today, high-quality wood is transformed into beams, planks and boards step by step, and attention is always paid to the grain, and the trunk turned if need be. When sawyers are assessing a trunk like this they say "they are reading the wood" (*ki wo yomu*).[10] The sawmills are highly specialized, many cut only a single kind of wood and supply a single trade. Thus for example there are sawmills for sliding door construction, for ceiling panels cut from the rootstock of Japanese cedars, for chests of drawers and packing cases in paulownia, for box-

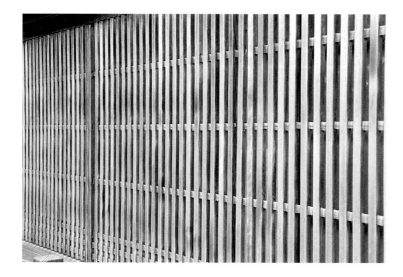

Lattice-work on a town house in Kyoto.

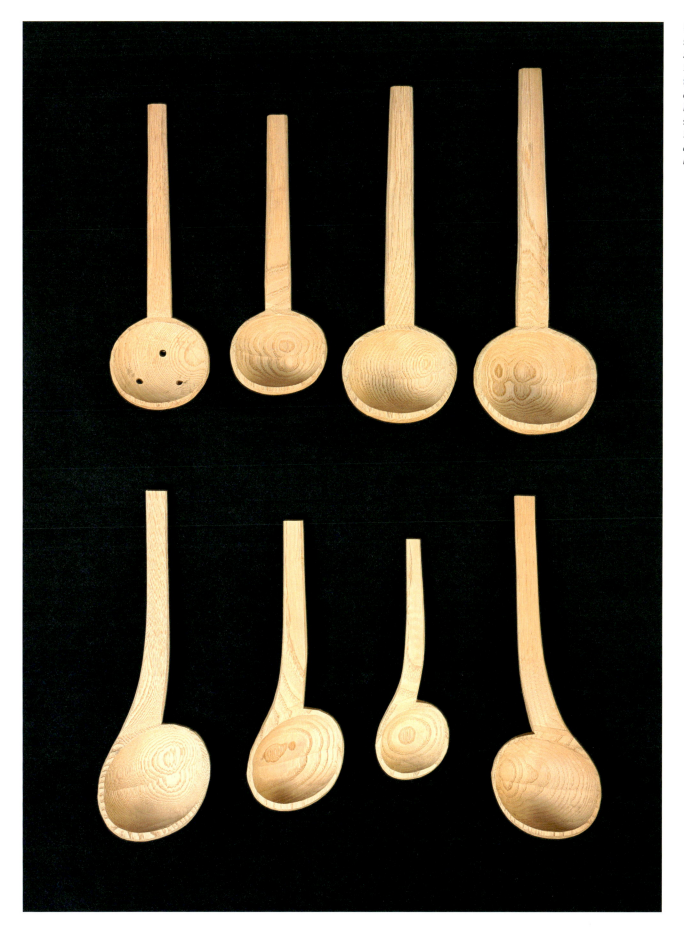

Chestnut wood spoons from the Atarashi workshop, Sotani. The spoons in the bottom row are asymmetrical, the handle is set to the side of the bowl. Spoons of this kind are carved in left- and right-handed versions.

wood combs, for tool handles and plane bodies in Kashi or for the specialized wood needed for building musical instruments. There are timber dealers for building and interior work who have specialized in high-quality structural timber, the *mei-boku-yu*, "dealers in famous woods." Here you find a lavish selection of posts and planks for picture alcoves or wide ceiling panels in Japanese cedar, for example.[11]

Prices in the wood trade differ extremely, which shows how highly certain woods and particularly attractive markings are esteemed. Attractive grain is often the most important criterion for the final price of a product. When conducting research about turners on the island of Miyajima off Hiroshima three small tea caddies were bought from a workshop. They were all made of local pine, were the same size and had been made in precisely the same way. And yet the turner charged quite different prices for them, 5000, 8000 and 16,000 yen. He had made the most expensive caddy from highly resinous rootstock, and when it was held against the light the wood was almost transparent. Something similar happens in the case of Japanese harps: here an instrument with markings reminiscent of pearls in many scattered concentric circles (*tama-moku*) commands almost four times the price of a comparable instrument with less elaborate grain. Japanese woodworkers used to use wood as if they were painters. Wood is particularly chosen for its effect at the places where it is most conspicuous. This is particularly clear in the case of temple and villa gates, for example. Craftsmen like to choose a highly attractive and lively texture for the panels, which are only a few millimetres thick, and not usually glued. When doors have two leaves, boards split in the middle are usually used, to give a symmetrical look. Where possible, all the panels are cut from a single trunk, to achieve visual uniformity. The same also applies to posts and beams, steps and ceiling panels. Wood is always chosen in terms of pictorial composition, and the wood is shown from its most attractive side. This sensitivity, which can be realized in almost all historical buildings and products, has now unfortunately become rare in Japan as well, because the preliminary working stages have been mechanized and hand-held machines are increasingly available.

Surfaces

The aesthetic effect of wood is further enhanced by the surface finish. Great attention is paid to the finish in almost all wood structures and products, to what can sometimes seem an excessive extent. This approach is promoted by the fact that the wood is usually not coloured, and so the material itself is the show side. Flat surfaces are usually smoothed with a plane. The shavings that are removed are often only 0.02 to 0.03 mm thick, the surface is mirror-smooth and glows slightly. Finish is the criterion by which a carpenter's or joiner's mastery is first assessed. Interestingly, building components and objects that were finished with adzes and planing knives before planes with fixed irons started to be widely used in the 15th century also show comparable care and astonishingly clean surfaces. Sanded surfaces are taboo in the building trades, but they do appear in some other wood crafts. Comb makers work the teeth with sticks with horsetail stuck to them, and when cherry bark is made into containers the surface is finally wet-polished with Tokusa leaves. These leaves are as rough as sand, and were also used by turners and box-makers until the introduction of sandpaper. Planing and polishing aim to produce the smoothest possible surface, but the opposite applies in many cases where a relief-like surface is sought. Soft woods like paulownia and coniferous timber are often brushed with willow roots when they are being made into chests of drawers, boxes and small, high-quality household goods. This rubs away or compresses some of the softer, early-growth wood, so that the lines of the harder late-growth wood stand out in a raised pattern. The grain is accentuated in this way, and the surface made easier to handle.

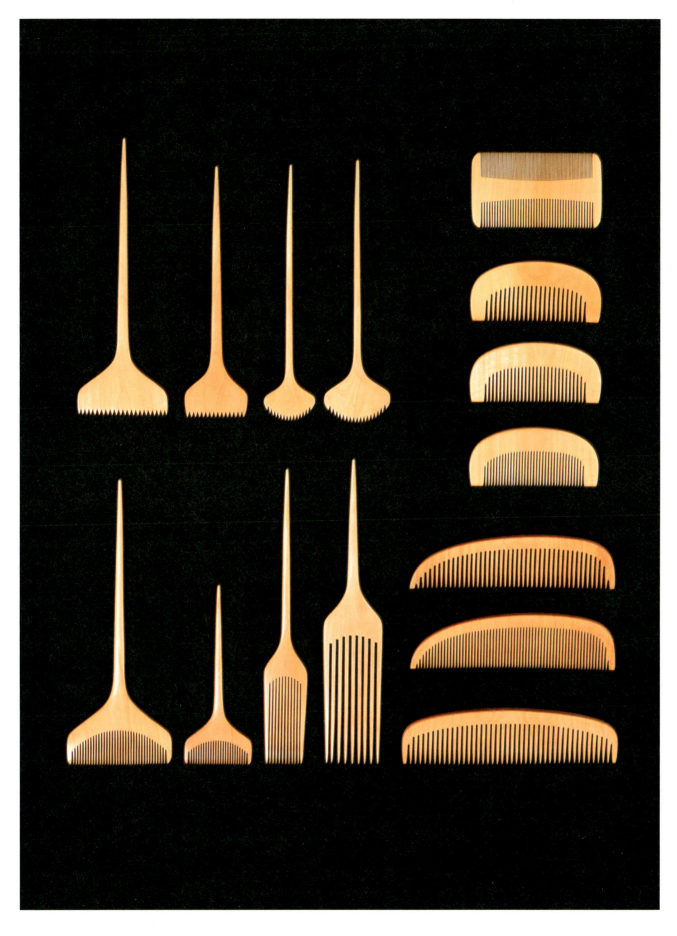

Boxwood combs from the Jûsanya workshop, Tokyo. The combs on the left with their long handles are used to pin the traditional coiffures, the combs on the right for cleaning hair.

Polished cedar trunks, as used for picture alcove posts.

Sometimes the characteristic traces left by certain tools are seen as desirable as well. Adze-finished planks are often used for the step between the entrance area of a house at ground level and the higher floor level. This appreciation of surfaces that seem natural and simple brings us to the tea cult, whose aesthetic ideals have greatly influenced house building since the 17th century. Here a whole number of "natural" or "unfinished" surfaces are preferred. Unshed round timbers are often used for the picture-alcove support or the head-tie beam. Trunks like this, with the bark left on, were also used for the gate supports in the garden of the Katsura villa near Kyoto. Round timbers – especially of young Japanese cedar – carefully debarked and then polished with sand (*migaki-maruta*, polished round timber), are a preferred choice for tea-rooms and houses.[12] These components may look natural and unassuming, but working them to such a perfect finish and fit is much more demanding and time-consuming than is the case with squared timbers. The surface of split wood is also seen as attractive, where the grain structure is particularly clear. It is used in suspended ceilings, for example, made of strips of wood that have been thinly split and then woven, on the under-side of wooden shingles or occasionally for containers for the tea ceremony.

Some apparently contradictory ideals of beauty occur in the signs of ageing. In Shintoism, newness is a much sought-after quality, as it is pure and unsullied. The shrines at Ise are still replaced every 20 years. In other places periodic renewal has been abandoned for reasons of expense, but the tradition has been continued on a reduced scale through repair campaigns.[13] Trays for sacrificial gifts and ritual utensils are also renewed frequently. Similar ideals of purity, which may also be driven by the humid climate, also made their mark on temple repair, for example, when a few centimetres were cut off the weathered ends of the rafters and the surfaces are planed down repeatedly.[14] But on the other hand there are examples in which traces of ageing are not just accepted, but sought after and even anticipated. Negoro lacquer work is well known; when the black upper coat of lacquer is rubbed away, the red ground appears under it. This effect, with red speckles appearing on the dark vessel, finally became so popular that it was created even on new vessels in the lacquer workshops.

Geta from the Shindô workshop, Chichiwa. Sandals in light Kiri wood. The men's version on the left is almost rectangular, with unobtrusive black straps, the women's version on the right is rounded and has straps with flower patterns. In the middle is a pair of children's sandals.

Geta with high hardwood bridges from the Shindô workshop, Chichiwa, the preferred footwear for cooks. The straps are made of buckskin.

25

Practising the profession, passing it on, documentation, funding
After a period of a good 100 years in which the country had disintegrated into many quasi-independent provinces, it was reunited step by step in the second half of the 16th century and placed under centralized rule. Society started to be divided into four classes, samurai, peasants, craftsmen and merchants (*shi-nô-kô-shô*). Social mobility was thus eliminated. The classes were divided physically as well, as they lived in their own quarters in the fortified cities. Craftsmen in one town came together in guild-like associations according to trade (*nakama or kumi*), this way protecting their market from outside competition and practising price and quality control. Professions were hereditary by law, so succession was automatically secured. These restrictions disappeared within a few years during the 1868 Meiji reformation, but they continued to have a mental and social impact for a long time. Professions continued to pass from father to son, and if there were no suitable male heirs a successor was adopted from the immediate family or the circle of former apprentices.

Apprenticeships were relatively long, lasting five years, and craftsmen often stayed with their masters for a few years out of gratitude. Until the post-war years apprentices usually lived in their masters' houses (*sumi-komi*). Training is still characterized by slow introduction to the profession and a lack of direct instruction. Apprentices spend their first years cleaning the workshop or sharpening tools. Constant repetition of the same task is seen as essential for acquiring techniques. Apprentices are expected to watch carefully as other craftsmen work and to learn their techniques "by looking". In Japanese this is called ude wo nusumu, "stealing skills". Asking the master directly for instructions would be interpreted as lack of attention.

Craftsmen were settled in most wood trades, and worked in a fixed workshop. Until the post-war years, carpenters did not have workshops. There were also some itinerant trades that mostly emerged as side occupations for farmers, like woodcutters, for example, the makers, called *kijiya*, of wooden cores for lacquered vessels, or spoon carvers. Most of them now have fixed workshops, the only exception being bark collectors and roofers specializing in bark, shingle and straw roofs.

Japan is as aware of the uniqueness of its crafts as it is that they are being lost. A nationwide inventory of all traditional professions, regardless of material, was drawn up between 1984 and 1933. A record was published for each prefecture. But all that is noted here are the most important data about the craftsmen; manufacturing steps are described only cursorily.[15]

Carved toys made of thin round timber from the Toda work-shop, Yonezawa. The lavish feather decoration on the hawks and chickens is made by a series of incisions.

Turned toys from the Niiyama workshop, Shirabu. Three Kokeshi dolls in the back row, in the front row three elaborately painted tops.

27

An act relating to the promotion of traditional crafts was passed in 1974, meaning that craft associations were entitled to funding.[16] Recognition as a traditional art or craft is based on various criteria: the use of traditional materials, manufacture largely by hand, and products intended mainly for everyday use. Only trades that have existed since the Edo period are eligible, and they must be practised by at least ten workshops or 30 craftsmen in a region. So it is not individual outstanding craftsmen who are supported, but centres. Four of the trades discussed in this book have been recognized in this way: house altars from Kyoto, trousseau cupboards from Kamo, coopers' goods from Odate and Japanese harps from Fukuyama. The seal of quality that is awarded, which has to be placed on the products, shows the sun from the Japanese flag under the stylized character for tsutaeru, "pass on". To secure a succession in the traditional professions and improve their acceptance in today's working world with its attested qualifications, craftsmen have since 1993 been able to take a "traditional craftsman's" examination when they have at least twelve years' experience in the profession.

Finally, since 1975 the Ministry of Culture has funded traditional techniques indispensable for the conservation of cultural properties. A particularly good craftsman or association can be recognized as the holder of a technique, and passing it on can be funded. This offers the possibility of protecting endangered crafts, especially in the field of monument preservation. Four of the crafts presented here receive such funding: monument preservation carpentry, architectural model-making, sliding door construction and wooden shingle roofing.[17]

The possibility of acquiring public funding that has been sketched out here has in many cases led to an increased awareness of the professions and to documentation of their techniques. But they will only survive in the long run as long as there is sufficient demand.

建築

建築

Architecture

Wooden bridges
Temple construction – restoration of a large hall
Workshop at the Great Shrine of Ise
Teahouses
House building
Architectural models
Wooden shingle and bark roofs

View of the Kintaikyô
bridge from below
of a bridge arch
reinforced with
cross braces between
the girders.

Wooden bridges

The appropriate Japanese character, read as *kyô* or *hash*i, indicates that bridges were first made of wood. Its left-hand half consists of the radical for tree, and the right-hand half of a character that taken on its own means "high and curved". If narrow valleys had to be crossed and piers could not be erected, arched bridges were constructed using the free cantilever method. For this several layers of cantilever beams were assembled on top of each other and joined at the apex (*gasshô-bashi*, *katamochi-bashi*, *mochiokuri-bashi*). The best-known example of this building method is the so-called Monkey Bridge in Ôtsuki, Yamanashi Prefecture.

Simple stilted structures were used for wide rivers; the horizontal members were supported by posts at short intervals, and two or more posts were usually linked by tie-beams and braces to form a rigid frame. The longitudinal joints in the horizontal members were placed over the posts. This most common bridge form is called *keta-bashi*; typical examples are the bridge over the river Uji near Kyoto (there are records of a first bridge as early as the mid 7th century) and the bridge at the Inner Shrine in Ise, also called Uji-bashi.[1]

The first stone bridge with individual masonry semi-circular arches in Japan was built in the port of Nagasaki in 1634 (Megane-bashi). This technique is said to have been brought to Japan by a Chinaman. But stone bridges were the great exception until the country was opened up. The few that were built usually had small spans and even then a building method similar to that for timber bridges was used most of the time. In the late 19th century, rapid industrialization and the construction of roads and railway lines brought bridge forms developed in the West to Japan, and timber was gradually displaced as a building material for bridges by steel, bricks and concrete.

View of the Kintaikyô bridge, on the hill in the background is the reconstructed castle tower.

Kintaikyô in Iwakuni

Iwakuni was built as a castle city for the province of the same name in the early 17th century. It is divided by the wide Nishiki river. The administrative centre was below the castle mound on the left bank, while craftsmen and merchants settled on the right bank. The desire for a permanent connection was realized in 1673 under the third feudal ruler Yoshikawa Hiroyoshi. It took the form of a five-arched timber bridge just under 200 metres long with intermediate stone piers, placed below the castle at a point where the river gets narrower after a long bend. The bridge was washed away the year after it was built, but immediately replaced, with some structural improvements. The large number of historical illustrations makes it clear that it was considered a technical masterpiece and a popular attraction.[2] Thanks to repairs carried out at short intervals, this second bridge survived for 276 years, until it was destroyed by floods in 1950. It was not the timber load-bearing structure that made it collapse: in fact the intermediate piers, built from large pieces of granite using the dry-stone method, had been undermined. Probably the elaborate riverbed paving upstream and downstream of the bridge had not been adequately maintained. The bridge, which became a city landmark as early as the Edo period (1603–1867), was rebuilt again immediately after this second loss. The superstructure was in timber, keeping faith with the original by reference to old plans, but the piers were given concrete foundations ten metres deep and a reinforced concrete core. The dry wall was then reconstructed around this, using the original stones.[3]

Coloured woodcut of Hirsoshige from a series showing well-known places from 6o provinces.

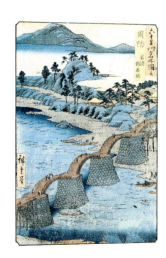

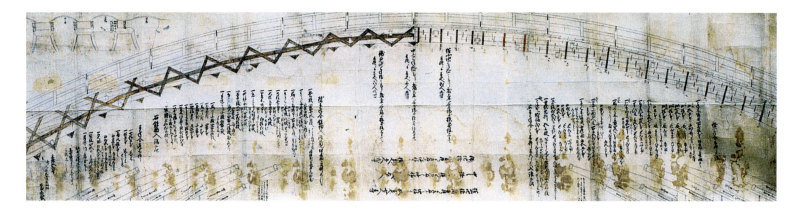

1699 plan of bridge arch; the left-hand side shows reinforcement with long braces and the V-shaped so-called saddle braces.

Like most historic wooden bridges in Japan the superstructure of the Kintaikyô was not roofed, like historical timber bridges in the Alpine region. It was thus open to the elements, and had to be completely renewed in three phases after another 50 years, between November 2001 and March 2004.[4] The work was carried out in the winter months each time, as precipitation is low.

One of this bridge's particular charms is that it combines the two commonest historical building methods. The three central arches with an inside width of 35.1 m are self-supporting and strongly arched; they were built using cantilever beams with a total of eleven layers. In contrast, the two outer approach spans are only slightly arched. Here the girders are supported by a total of five rigid frames each made up of three pillars connected by penetrating tie-beams and cross-braces. The feet of these supports are striking: they were made of granite to prevent rotting, and are connected with the timber supports by a straight hooked halving joint and flat iron bars, a method also used in traditional Japanese timber architecture for shoeing damaged supports. On all five

Streamlined bridge pier; the river-bed is reinforced with large granite rocks.

arches the core of the load-bearing structure consists of five parallel girders with about a metre between the axes. The emergence of one of the central bridge arches gives the best insight into the construction.

Bridge pier – some blocks are secured at the top with inset dovetails. Here a timber jointing technique has been applied to stone.

A bridge arch is born

The streamlined piers taper slightly towards the top, and here particularly one is struck by a number of stones linked by inserted double dovetails in a lead alloy. At the top of the pier, a recess serves as a bearing for the horizontal members. Steel sections have been in position here since 1952 to restrain the four lowest beam courses to the recess. The first beam course is staggered slightly at the foot against a high sill at the pier recess. At the top end a transverse beam (*hana-bari*) is tenoned on and secured with wedges. The second beam course protrudes a good metre further. The courses are fixed among each other with wooden dowels and their ends are tenoned into a transverse beam. The next beam courses are laid next to, but not flush with the transverse beams on the level below. Wedge-shaped filler timbers are fitted to some of the lower ends of the beam courses, to reduce the tilt on the beams above. From the fifth course onwards the ends of the beams are no longer restrained to the intermediate piers but linked in each case with a so-called rear transverse beam (*ato-bari*) by a dovetailed lap joint.

Once nine courses have been fitted, the two cantilever beams are closed to make an arch by inserting the "large ridge timber" (*ômunagi*). This is followed by the "small ridge

Fitting the first cantilever beam courses.

Mortising a transverse beam to the tops of the cantilever beams – the tenons are secured with wedges.

The arch grows gradually – the dowels, square in cross-section, intended to take the next course, can be seen on top of the beams.

Securing the beams by wrapping flat iron round them – work-men turning the flat iron over.

timber", above which the ridge beam (*muna-bari*) is lap-dovetailed, before finally the eleventh beam course is laid on the prepared hard timber dowels, which always have a square cross-section in traditional Japanese timber construction (this is probably because large drills did not appear until a comparatively late date). The horizontal members, packed on top of each other in courses and secured against longitudinal shifting by dowels and lap-dovetailed transverse beams, are held together – in a total of 240 positions – with iron bands (*makigane*). At the same time the stability of the beams placed on top of each other is increased by hammering in dogs, 3500 per arch.

The upper extremity of the horizontal members has so far been stepped. So that they can start to arch, wedge-shaped filler timbers (*ato-tsumegi*, literally subsequent filler timbers) are placed in position and also fixed with dogs. The "adjustment timbers" (*heikin-gi*) are now fixed on the filler timbers, or the eleventh beam course; the incisions for the plank covering can be seen on top of them.

The five bridge girders are now elaborately reinforced, individually and in relation to each other. To do this, long braces are first nailed to the two sides of the girders. Then short, V-shaped braces (*kuragi*) are applied, their legs crossing in the upper third. The connection is secured at the foot of the braces by an iron sheet, and at the top they extend to the top edge of the girders.

The overall stability of the core structure, which is made up of five parallel girders is increased by installing cross-braces in the fields formed by the girders and the so-called rear beams. They are called *furi-domi-gi*, which indicates that they are intended to avoid vibration.

Covering the scaffolding and view of the arch from below.

Cross-braces between the girders.

The top layer of the wooden girders are stepped.

Before the transverse planking is laid, the upper side of the beams is covered with sheet copper, to protect them against water seepage. On the two steeper sides of the arch the covering consists first of all of 30 low steps with a slightly trapezoid cross-section.

Fastening the planks.

The planks are slightly trapezoid in cross-section. A tongue is cut at the top edge to fit into a corresponding groove in the next plank up. Here the edges at the end of the boards are bevelled with a chisel.

Boarding up the outer bridge girders and covering the beam heads.

Assembling the rail starts with laying the cushion timbers.

Rail post with sheet copper covering.

Wooden bridges

At the top rear edge of the planks a tongue is cut to fit into a corresponding groove in the step above (*hana-kasane-bari*). In the top third of the arch the planks are laid flat against each other with a small gap between them, and linked by tongues.

The outsides of the two outer girders are finally clad in Hinoki timber boards. Vertical battens form the support, they have recesses so that the boards overlap slightly.

The many transverse beams that are set not quite flush with the bridge girders, they protrude sideways from the two outer girders by about 20 cm. To protect them from weathering, these beam ends are covered with copper sheeting and then finally with short boards. These coverings (*hari-hana-kakushi*, literally beam end covering) look like decoration lending rhythm when the bridge arches are finally shuttered, but in the first place they are merely functional.

Finally the carpenters turn to the rails. Their bases are not set directly on the bridge sur-face, but rest on the so-called "cushion timbers" mounted at 22 points per arch. Now the posts can be stuck into the prepared slits, a penetrating tie-beam (*tôshi-nuki*) pushed in and the handrail (*kasagi*, literally "umbrella timber") mortised on the top.

A special effort is made with the end posts (*oya-bashira*), which are much thicker; they are placed on a bevelled base and have a prism-shaped covering clad in sheet copper at the top. The posts and the top transverse timber of the rail are secured with flat iron bars.

Bridge construction site

The project's logistical demands alone are enormous. The timber for the renovation of the bridge – almost 500 m³ – was originally to be sourced from the region, but finally it had to come from all over the country, through a timber dealer who specialized in large-scale projects like this. Purchasing the timber swallowed up about half the total budget of roughly € 40 million. The total of 30,000 nails and 16,000 dogs – a single span with an overall weight of 60 tons contains about four tons of iron – were specially forged.

The timber work was carried out by 22 local carpenters directed by Master Ebisaki, who lives just by the bridge; his father was in charge of rebuilding the bridge 50 years ago.

The bridge arches were first drawn on the original scale in a workshop a good kilometre upstream, and plywood patterns prepared for all the important parts. To avoid wasting time on corrective work during the final assembly process, the supports were constructed once as a test (*kari-gumi*). This used to be done right by the river, now appropriate steel sections are available for this purpose in the area outside the workshop.

Originally the principal motif for cantilever beam constructions had been that they could be built unsupported without elaborate scaffolding. But boarding structures seem to have been used at an early stage for building and repairing the Kintai Bridge. The oldest surviving plan of 1699 contains a timber list that ends with thin round timbers, planks and ropes for building such a structure. Until the 1953 renewal the scaffolding, which was also used as a working platform and for adjusting the beams was lashed together with round timbers, but today tubular steel is used, for safety reasons.

The charm of the building site lies in the applied construction and the high-calibre craftsmanship. Only the complete replacement gives food for thought. Actually it is a break with tradition, as until it was lost in 1950 the bridge was only repaired. Demolishing it showed clearly that apart from areas exposed to wear like the surface covering, the rails and the external boarding, the actual load-bearing structure was still in very good condition. It would have been more economical to replace damaged sections only; this would also have guaranteed that bridge-building techniques were passed down to local carpenters. The master carpenter points out that in 50 years there will probably not be a single carpenter now involved who would be capable of carrying out a renovation.

Bridge construction site – the river is diverted with sandbags for the duration of the building work.

Temple construction – restoration of a large hall

In Japan like in most other countries, the greatest possible effort in building was focused on places of worship, and especially Buddhist temples. They are built of the best timber, constructed to the highest standards, well cared for and seldom converted to other uses, which makes them the most long-lasting of all the numerous Japanese architectural types by a long way.[5]

Of all the Japanese carpentry professions, the *miya-daiku*, or shrine carpenter, who specializes in building places of worship, certainly enjoys the greatest prestige. New temples and shrines are still built in Japan, often based on historical models. And there are isolated cases of complexes that were destroyed in the Second World War, or even centuries ago, being rebuilt; a particularly well-known example is the Yakushi-ji temple near Nara. But alongside new construction, the *miya-daiku* was always responsible for maintaining and repairing places of worship, many of which are listed as ancient monuments today. Ancient monuments have been subject to public preservation in Japan since 1897, when a law was passed to protect old shrines and temples. The range of listed architecture has now gradually increased to include castles, residential buildings and early architecture influenced by the West, and industrial monuments, but even now places of worship make up about two thirds of the monument stock. Important features

of temple construction, and also of the restoration of such buildings, will be presented here, taking the great Main Hall of the Shôkô-ji as an example; it was comprehensively restored from 1999 to 2004.[6]

Main Hall of the Shôkô-ji

The temple stands on a plateau above the Fushiki Bay on the Sea of Japan, about 270 km north-west of Tokyo. The temple complex, which has survived largely in its early 19th century form, and was originally surrounded by a moat, measures 150 m from east to west and 180 m from north to south. The northern half contains the residential and domestic buildings; the dominant Main Hall and other buildings used for ritual like the two-storey bell-tower or the Sutra hall are in the southern half.[7]

The Main Hall was built between 1793 and 1795. With a width of almost 40 m and a depth of 37.5 m it is one of the largest temple halls in the country. The ridge of the enormous hip-and-gable roof rises to a height of 23.5 m. Like most main temple halls of the Pure Land School, the hall of the Shôkô-ji faces east; the Buddha Amida, the main image in these halls, sits in the west according to the scriptures, facing east to greet the faithful in his paradise. Pilgrims reach the hall via a wide, roofed staircase on the façade side (*kohai*) leading up to floor level, which is a good two metres higher. The interior is divided into two areas by doors and transoms with open-work carvings. At the front is a large room with tatami mats for the congregation, with a U-shaped board-floored open veranda running round it. Behind the screen in the rear third of the hall, is the central inner sanctuary (*naijin*) with the altar and a shrine placed on it, housing a standing fig-ure of the Buddha Amida. To the right and left of the inner sanctuary are side-rooms (*yoma*), also with tatami mats. At the back of the hall is the two-bay deep rear hall (*ushi-ro-dô*), running across the full width of the building. The congregation attended the rit-uals in the outer sanctuary, the area behind the elaborately designed partition was reserved for the priests of the Shôkô-ji and members of its subsidiary temples.

Like all places of worship dating from after 1300, a modular system was used when planning this hall as well.[8] Here the distance between the column axes is an exact mul-tiple of the distance between the rafters. The columns are circular in cross-section in the

View of the outer sanctuary; in the foreground, a large lantern is suspended from the coffered ceiling.

inner part of the hall, otherwise square, and rest on individual foundations. They are linked together on up to six levels by penetrating and halved tie-beams. Most of the column tops support bracket complexes, a feature synonymous with Buddhist temples in Japan. Above this, long, thick round beams are placed longitudinally and transversely, and the roof structure is set on them. The thinly dimensioned supports of the roof structure, connected to each other by penetrating tie-beams, do not run through to the purlins, but just to the next layer of beams, thereby giving this high roof frame a four-layer structure. In contrast with medieval buildings, the roof supports do not relate directly to the position of the columns. The roof diagonals feature the typical concave curve, topped by a striking box-shaped ridge (*hako-mune*). The lavish decoration is typical of the building period. It shows particularly clearly on the relief carvings on the beams and the lavishly adorned roof pediments.

Acquiring the building materials for this massive hall must have created difficulties. A good quarter of the total of 122 columns were elaborately shod to achieve lengths of up to 9 m. There are also some unusual columns for which the foot of the trunk was used, rather than being cut off. In the inner sanctuary in particular, where the columns are painted in black Urushi lacquer or gilded, timber species like chestnut and cherry were used; these are not suitable as building material because they are susceptible to damage.

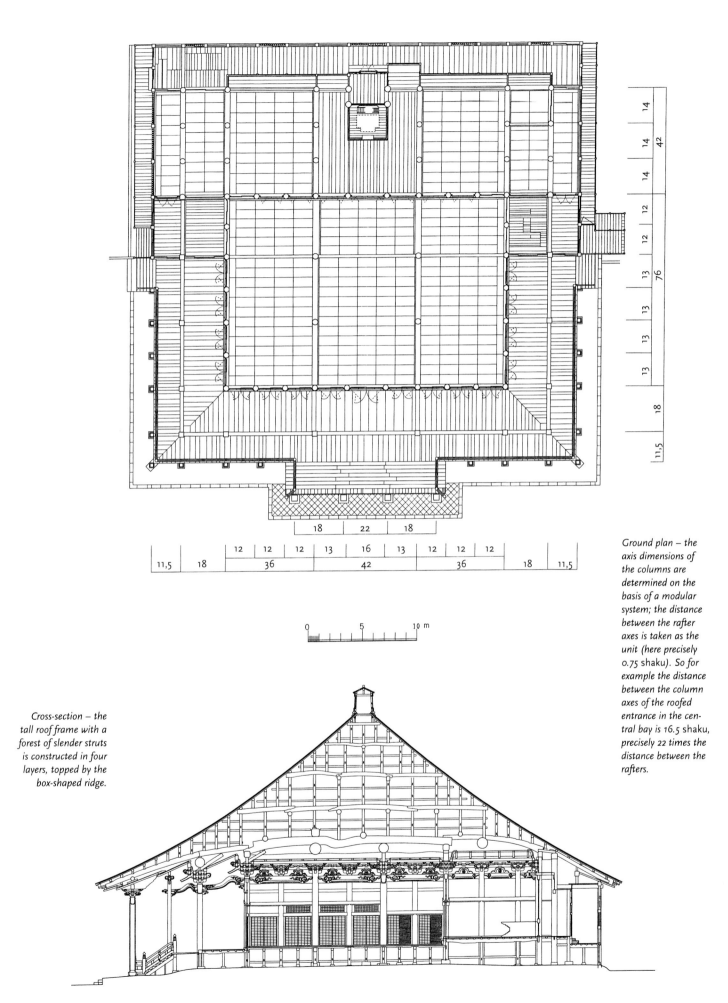

14

14   42

14

12

12

13   76

13

13

13

18

11,5

18   22   18

11,5 | 18 | 12 | 12 | 12 | 13 | 16 | 13 | 12 | 12 | 12 | 18 | 11,5

36   42   36

0   5   10 m

Cross-section – the tall roof frame with a forest of slender struts is constructed in four layers, topped by the box-shaped ridge.

Ground plan – the axis dimensions of the columns are determined on the basis of a modular system; the distance between the rafter axes is taken as the unit (here precisely 0.75 shaku). So for example the distance between the column axes of the roofed entrance in the central bay is 16.5 shaku, precisely 22 times the distance between the rafters.

Temple construction

The temple was supported by the Daimyo of this province until the late Edo period (1868), but got into economic difficulties after this quasi-public funding was withdrawn. Another disadvantage was that it had only a very small congregation, and no burial facilities, which is one of the principal sources of income for Japanese temples today. Maintenance of the temple deteriorated as a consequence, and the building was actually unoccupied for a time after the death of the last high priest. Finally in 1988 first the Main Hall, followed by eleven other buildings in 1995, were listed as "Important Cultural Properties". In the meantime the eaves had partially started to drop, and some areas were in danger of collapse. After protracted negotiations had secured finance for the restoration programme (public funding by the state, the prefecture and the community reached the highest possible level of 97 %), restoration began in March 1999.

Restoration

In Japan, extensive restoration always starts with complete enclosure of the building, in this case using steel scaffolding. This makes the work independent of the weather, protects the building from the elements during the programme and also provides "storage" for dismantled sections. As a rule the listed buildings are partially or even completely dismantled. The West is often critical of this practice, but does not take account of the fact that the framework of Japanese timber constructions is made up of a large number of components simply fitted together and secured with wedges, and so like a building kit it can be dismantled and rebuilt almost without damage.

A detailed damage assessment showed that the columns had settled by a maximum of 91 mm, and the entire hall had tilted to the east. With a column length of 7.28 m the tilt reached a maximum of 149 mm. 15 foundation stones for the columns and veranda struts had split. Damp rot and a high level of insect damage was discovered in column feet and parts of the floor substructure. The most alarming damage was to the lower courses of the roof beams and the cantilever beams securing the eaves. Some of them were already broken, causing point settlements of up to 40 cm.

The carpenters working under Master Tanaka started by dismantling the entire interior, including doors, ceiling panels and also the floorboards. The roof covering and the rafters underneath it were also removed. The carefully dismantled parts were wrapped in some cases and stored in the enclosure or in specially constructed sheds. The parts had to be carefully marked to identify them for re-assembly. The carpenters used jointing marks of the kind still employed for this purpose when building new halls and traditional houses. The marks ensure that the position of each column is determined by two co-ordinates. Little pieces of plywood are pinned to all the pieces that have been removed and the co-ordinates of the next column, the name of the component and its orientation are noted on them. The care taken for this partial dismantlement was not restricted to high-calibre interior elements, but extended to the nails and dogs. The examination was carried out equally meticulously. For example, the wooden shingles under the pantile roofing were removed individually, by hand. This effort was rewarded because over a 100 building inscriptions were found, most of them dated, thus providing a fairly precise account of roof maintenance work.

In contrast with Europe, for example, cost considerations play a minor role at best when restoring the relatively few state-protected buildings in Japan, but as this hall was so big, the decision was taken not to dismantle completely, also to save costs. So all the roof components had to have their load-bearing capacity carefully assessed. This used to be done just by tapping the beams with a little hammer: the sound then indicated the extent of the damage. But here a resistograph was used on a Japanese monument site for the first time. Resistance to drilling was measured at each of four positions at both ends of the 41 selected beams in the roof and the results recorded in diagrams. The investigations proceeded smoothly and caused little damage, as the drill-holes were only 3 mm in diameter.

Correcting the deformations

The hall had tilted eastwards by about 15 cm. In Europe this deviation, which gave no cause for alarm statically, would have been deliberately retained as a historical statement. Things are different in Japan: correction is taken for granted. There are two particular reasons for this attitude. The load-bearing frame functions without triangular bracing; deformations do frequently occur, but they are relatively easy to deal with because the joints are largely only pegged and wedged. In highly precise buildings with an orthogonal structure of posts and rails, and also a large number of sliding elements, the deformations impinge much more, and are also aesthetically disturbing.

If the building is completely dismantled, the deformations are relatively easy to correct when it is re-assembled. Here it was decided that the building should be corrected as it stood, which is more difficult, and occurs less frequently. First the carpenters removed all the halved tie beams, sills and lintels running east-west and uncovered a strip about 3 cm wide on the few wattle-and-daub infill panels in the rear part of the hall. This made the structure considerably more supple and pliable. Then the settling was provisionally corrected by raising the columns and inserting wooden packing under their feet, which brought the building back on to an even keel. Now double-T girders could be attached to the columns running north-south and east-west, and the hall jacked up by a total of 15 cm, in three stages. Minor displacements of the column feet were corrected with cables, to bring them back into alignment.

Finally, the tilt on the hall was corrected. To return the hall to its original state, the heads of six rows of columns were attached to the beams above by 16 mm steel cables. Then a total of twelve horizontal steel cables were attached to a steel girder that had previously been fixed to the back of the large steel enclosure. Then, in late May 2000, chain blocks pulled the hall frame very gradually westwards in two days. The deformation correction lasted for about a year until the roof boarding could be fitted. The hall was deliberately ""over-pulled" by about 1-2 cm, because it was assumed that the frame would deform slightly again when the cables were removed.

Once the hall was standing plumb again, cracked foundation stones were replaced and attention turned to repairing the column feet, a notoriously weak feature of many historical timber buildings.

Replacing the column feet

Of the 122 columns in the hall, 54 were packed and 22 provided with a new foot. Sheet lead was used to pack up to 15 mm, as it adapts well to the unevenness of the foundation stones. Blocks of Keyaki wood were used for corrections and repairs up to a height of 80 % of the column cross-section. Longer replacements were shod in the direction of the grain, and jointed with a cruciform tongue or a scarf joint. New feet rising above floor level were required for two columns in the rear of the hall. These had to be fitted in situ, and a particular sequence had to be followed when fitting the shoes because of the penetrating tie-beams.

Because of lack of materials, some timber species had been used when constructing the large hall that are actually not suitable for building. But the original timbers were faithfully used for the repairs as well, so a cherry and a cedar column were repaired using precisely this timber. Master carpenter Tanaka was not entirely happy about this, as Keyaki wood replacements would have been much more durable. Conservation architect Imai is well aware that these replacements are not suitably durable, but sees the choice of building material as a historical statement that he wants to preserve in the restoration work.

*Typical new footings
–"shoes"– for the
columns.*

*Long replacement
shoe with scarf joint
and tongues.*

*Fitting the replace-
ment shoes in situ.
The grooves on the
foot of the columns
allow condensation
to run off.*

The new column feet and also the hardwood blocks that had been put in position had to conform precisely to the surface of the uncut field-stones used as foundations for the columns. Field-stones occur in some of the oldest surviving wooden buildings in Japan, for example at the well-known Shôsôin log cabin-style storehouse at the Tôdai-ji temple near Nara. The precise adaptation of the column feet to the foundation stones (*hikari-tsuke*) is achieved by working them to correspond with the irregularities in the stones. This is an entirely different approach from the Western one, where the foundation stone would be levelled to deal with a problem of this kind.

Because the columns in this enormous hall have such a large cross-section, the master carpenter devised a new process. He first made a plaster cast of the foundation stones, having covered them with plastic sheeting. Then he drew an ink grid on the cast, with lines roughly 4 cm apart. Profiles were taken with a comb (*mako*) at all the co-ordinates and cardboard templates cut accordingly. The same grid was then applied to the new feet, and they were worked to the precise dimensions with chisels and small curved hand planes. Columns, feet and foundation stones were marked with centre lines (*shin-zumi*) so that their position could be determined easily and precisely even when the cross-sections were irregular. These centre markings incidentally are essential for the use of round timbers for frameworks and roof structures, which is widespread in Japanese timber construction. In Europe timbers are always marked out from the marking side, that is to say from the edge of a component that has already been squared up or hewn. So it is not surprising that round timbers have played a minor part in histor-

46

ical European timber construction, except in log cabin-style structures and simple temporary buildings.

After the correction process, replacement of cracked foundation stones and repairs to the column feet, the 600 ton structural framework of the hall was lowered back on to its foundation stones in January 2001, before turning to repairs of the roof structure.

Repairing the eaves and roof frame

The resistograph measurements had confirmed the original impression that several roof beams and a large number of cantilever beams for reducing strain on the eaves had

been so badly damaged by insects that they had to be replaced or the load on them eased. To replace 13 of the total of 70 cantilever beams along the eaves, the roof frame was partly removed in the area above them. The beams that were to be replaced were cut with a chain saw to make it easier to take them out. The roof struts and purlins could return to their original position after the new cantilever beams, which were in stripped pine, had been installed.

It proved much more difficult and time-consuming to replace three roof beams and fit one suspender beam. They form part of the beam courses supporting the roof frame, which is over 15 m high. To change the three beams, the struts standing on them were suspended from the next higher layer of beams, using steel cables and braces temporarily fixed with nails. Then the old beam was cut and taken out. The struts above the beam that was to be replaced were now shortened by a few centimetres at the bottom end and a new short tenon was cut. The new beam, which had been rolled over a makeshift path made of planks into the roof frame, could now be pushed into position from the side. Finally, two pieces of hard Keyaki wood enclosing the short tenon at the bottom were fitted under the struts and fixed on to the new beam with screws.

The strain on a weakened roof beam above the outer sanctuary was reduced by installing a suspender beam. This was similar to the procedure that had been used for the 1905 repairs. A pine trunk, squared on two sides, was used to make the beam, 11.7 m long and weighing almost five tons. It was lifted into the enclosure on one of the narrow sides of the hall with a crane, and then pushed on round timbers into the roof frame of the hall at the pediment. The beam, on which all the carpenters involved had written their names with an ink brush, fitted precisely and fell into the prepared recesses of its own accord. Several scaffolders helped to replace the extremely heavy pine trunks, as the procedure presented a considerable risk to both the workmen and the hall.

Interior work and local repairs

After the hall's framework and roofing had been refurbished, attention could turn to the interior and to local damage. To preserve as much of the historical substance, and thus as much information, as possible, repairs are preferred to replacing parts of the building, even if they cost more time and money. Thus for example the damaged bottom transverse member was replaced on some of the sliding doors. Wide cracks and damaged areas were filled with pieces of wood that had been carefully shaped to ensure a precise fit. The same material, and also similar grain patterns, should be selected to fill, lengthen or widen in this way. Such care is not restricted to visible sections: parts were

Intricate repairs at the top of the box-ridge. This plank will subsequently be protected with sheet metal.

thoroughly repaired in the roof frame and also in the box ridge, which was subsequently covered with sheet copper again.

The bottom rail on the sliding door was renewed with a gooseneck mortise-and-tenon joint. Heavy sliding doors like this one often have a wooden wheel built in at the bottom, so that the door runs on wheels.

The master carpenter

The timber work on the hall was directed by Master Tanaka, who put together a team of ten carpenters for the job. Tanaka is a third generation master carpenter, and comes from the neighbouring town of Takaoka. His father and grandfather built traditional houses, but at the age of 36 he gave up the family business that he had taken over only a few years before and worked for a well-known temple-builder for ten years. He had been captivated by the temple roofs with their elaborate bracket systems and their elegant curves. Then Tanaka started to work independently again in the year 2000, to take over the Main Hall restoration project. He called his workshop Tanaka-kôshô; kôshô is an old word for carpenter, and Tanaka intended it to have deliberately traditional connotations. He sees the relationship with materials as the thing that makes a restoration site most different from a normal construction site. As a carpenter, he says he is used to choosing timber according to his preferences, but on the monument building site he has to hold back – in Tanaka's own words, he has to throw his ego away. Subordination is what is called for. The carpenter has to think himself into the building, respect it, and intervene in it as little as possible. For example, he points out that when making a replacement everything is already decided, the type of wood, dimensions and the quality of the wood; he has very little scope left. Of course there are places in the building where he would act very differently if it was new. But ideas of this kind are not permissible on a monument building site.

The fact that Tanaka fundamentally sees himself as a kind of successor to the old carpenters also shows in the way he treats his apprentice. The apprentice is now 23 years old, and lives with his master. This used to be taken for granted, but it very rarely happens today.

Tanaka carving a fitting piece to be inserted at a damaged area on one of the veranda rainbow beams.

During the repair work, the hall was protected by a complete enclosure.

Workshop at the Great Shrine of Ise

Ise is in the east of the Kii peninsula, a few kilometres from the sea, at the foot of the mountains. Here, about four kilometres apart, are two central Shinto shrines, the Kotai-jingu or Inner Shrine (Naiku) and the Toyokedai-jingu or Outer Shrine (Geku), and also a total of 90 minor shrines of varying degrees (14 *bekku*, 109 *sessha*, *massha* and *sokansha*). The deity Amaterasu Omikami is worshipped in the Inner Shrine, sun goddess and also ancestor of the Japanese ruling house. The main building houses the mirror, symbol of the sun goddess, and one of three regalities. The Outer Shrine is dedicated to Toyouke Omikami, the deity of food, clothes and housing. As the central shrine of Shintoism, Ise has been a place of pilgrimage from the Middle Ages to the present day.

The main buildings in the various complexes, all facing south, are enclosed by wooden fences and not accessible; there are even four staggered fences at the Inner and the Outer Shrine. The main and minor shrine buildings are constructed in the same way, and share their most important design features. They are all post structures with high floors, and the bays between the horizontal and vertical members are filled with boards set in grooves. The buildings have gabled roofs sloping at an angle of 45°, with the entrance located at the eaves side. The roofs of the Outer and Inner Shrine are covered with reeds (*kaya-buki*), while those of the minor shrines are covered with boards (*ita-buki*). The unpainted cypress wood soon takes on a silvery-grey colour.[9]

There are two building plots for each of the main and the minor shrines. The structures are rebuilt at 20 year intervals, a new shrine is set up, the images are relocated in a nocturnal ceremony and the old shrine is taken down a little later. This is called *shikinen-sengu* in Japanese, literally translated, moving a shrine in festival years. It implies renew-

Aerial view of the Inner Shrine immediately after the new complex was completed in summer 1995. The old and new shrines stand next to each other for a few months.

The minor Aramatsuri-no-miya shrine immediately north of the Inner Shrine. Here the two neighbouring building plots can be clearly made out; the steps on the left-hand side have been built around a large cypress.

Small minor shrine, showing the old and the new shrine. The old trees have been left intact, and the site has only been minimally levelled, as can be seen from small differences in level between the two sites.

ing a shrine complex at certain intervals. Ise, including its minor shrines, is now the only place in Japan where complete physical renewal still takes place.[10]

Shikinen-sengu – periodic renewal of the complex

Many Shinto shrines used to be periodically renewed in this way. The intervals differed: at Izumo-taisha it was 60 years, at Kitano-jinja 50 and at Sumiyoshi-taisha every 20 years. The Ise complex was first rebuilt in the late 7th century. With the exception of a 120 year interruption between 1465 and 1585, when the country had broken down into warring provinces, and some delays as at the end of the Second World War, the complex was regularly renewed. Renewal was abandoned at the other shrines mainly

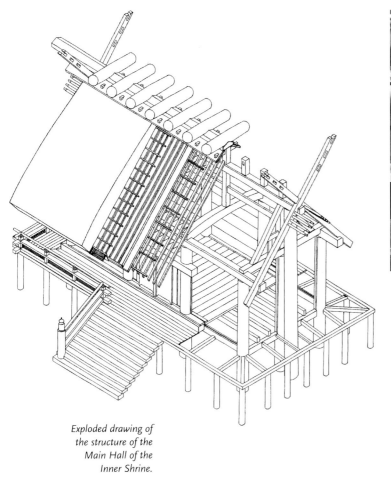

*Exploded drawing of
the structure of the
Main Hall of the
Inner Shrine.*

*Model of the Main
Hall (Goshhô-den)
of the Inner Shrine
(Kotai-gingu),
scale 1:20.*

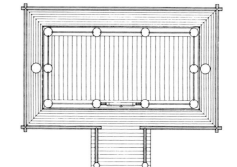

*Views and ground plan of
the Main Hall of the
Inner Shrine.*

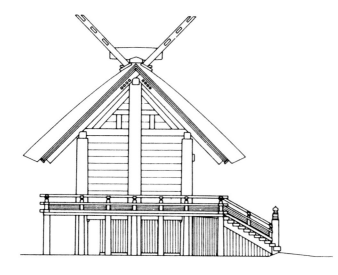

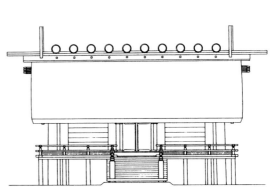

on grounds of cost, but it is often practised in reduced form or as pure ritual. At the Kasuga-taisha in Nara, for example, the buildings are repaired every 20 years: the red finish is restored and the ritual objects are replaced.

The motives for this elaborate rebuilding are complex. Renewal and purification are central themes in the Shinto religion. The sacred objects symbolizing the deities are to be accorded an immaculate home. Additionally, the buildings, with their posts fixed directly in the soil and the roofs covered in reeds or boards only last for a very limited period.

The rhythmically repeated building campaigns meant that archetypes of Japanese architecture have been faithfully handed down here. The shrines' construction and dimensions reflect 7th century architectural forms, which have also come down through the Haniwa figures on grave mounds and through depictions on ritual bells. Structurally the buildings are very similar to the rice storehouses that were customary at the time, and the elaborate decoration on the ridge created by the barrel-shaped timbers (*katsuogi*) placed on them and the extended boards at the edge of the gable wall (*shigi*) go back to the houses of the nobility. Another consequence of the periodical renewal of buildings and of devotional items is that the craft skills are handed down to subsequent generations unchanged.

In 1868 the military Tokugawa regime ended, the Tenno shifted back into the centre of the political system (Meiji restoration) and Shintoism became the state religion. Shortly afterwards, in 1887, an office of shrine-building was set up in the Ministry of the Interior. The periodic renewal of the Ise shrines thus became a public building project, which also resulted in the keeping of systematic records. Plans of all the buildings were drawn, and precise lists of timber and of the trunks needed compiled. But some of the details were changed, for example a sheet copper sheath was applied under the reed roof to prevent damage to the treasures inside.

Workshops at the Great Shrine

Prior to the 58th Sengu, which was completed in 1929, permanent workshops with ponds, sawmill and above all sheds for storing and drying the building wood, about 40 buildings in all, were set up on a site of about 8.8 hectares by the Outer Shrine. The Yamada workshops are run by the Great Shrine and were named after the land in which they stand. The timber for renewing the two large shrines and for the 90 minor shrines in the region is still cut, dried and worked here. As well as the sawmill, there are two workshops each with four carpenters, a sheet metal workshop for the numerous copper coverings and also a workshop preparing bundled reeds for the roofs. The main shrine at Ise is probably the only place in Japan where everything is controlled by a single organization, from forestry via felling and cutting to construction. There are also few places in Japan where wood is worked with such expertise and empathy.

The way of the wood

About 11,000 m^3 of wood are needed to renew Inner and Outer Shrine complexes with all their side buildings and for the many minor shrines. About a third of the trunks have to be over 60 cm in diameter, so that large posts or purlins can be cut in one piece from one trunk. The posts supporting the ridge purlin in the central shrine are the largest components, with a diameter of 76 cm and a length of 11 m.

Until the Middle Ages, wood for renewing the shrine came from the immediate vicinity. Since the 13th century it had to be sourced from increasingly distant regions, and from about 1650 to the present day it has come from the mountainous Kiso area in the Nagano Prefecture, which is famous for its cypresses. In the early 20th century a woodland area of 8000 hectares with particularly old trees was reserved for renewing the shrines in Ise, and put directly under its control. After the end of the war Shintoism was no longer the state religion, the Great Shrine's woods were nationalized and the timber had to be bought in from then on; in 1996 felling the state forest cypresses, which were several 100 years old, was greatly restricted, as an end to the stock was anticipated. This restriction caused considerable price increases: between the 1973 and 1993 renewals the price per cubic metre had increased almost tenfold, to € 5000 (total cost over € 50 million).

The end of a trunk –
the South Gate of
the Inner Shrine will
be covered by the
three middle planks
to be cut from this
trunk.

Marking out always
starts with the centre
line. Here it is being
worked out by Okada,
the master responsible
for the most recent
renewal of the
shrine.

For the coming 63rd Sengu in 2013, supplies will again be drawn from an area of 2900 ha in the upper reaches of the nearby Isuzu river, which the main shrine had afforested in the 1920s. The trunks will be a good 70 years old by then, and should cover 20% of requirements. Care and felling of the forest properties, which purchases have now brought up to almost 7000 hectares, are in the hands of the shrine's own forestry department, which has 20 employees. And so after several centuries it will be possible to source some of the timber locally. The intention is to increase this gradually, but it will take at least 130 years before the shrine will be able to provide its own timber again. Until then the larger components will no longer be made of cypress (Hinoki, Chamaecyparis obtusa), but of Hiba, which comes from the north Japanese prefecture of Aomori. It is closely related to the cypress, can be worked just as well, rots more slowly and is more resistant to insect attack. It does have a sharp and pungent smell, however, and inclines to shrinkage cracks. The price of Hiba is only a third to a quarter of that of Hinoki.

The trees are felled in the winter months from October to February, and delivered to Ise by lorry. An inventory number is carved at the top end on arrival. Before being cut, the trunks are stored in ponds for up to three years. This not only avoids shrinkage cracks, but apparently it also draws the sap out of the timber, which then dries more quickly after milling.

Marking out

The trunks are first cut into lengths in the sawmill as needed. Then the master carpenter turns them to check for damage and growth defects. He works on the top ends of the trunks with electric and hand planes, as it is easier to mark out on a smooth surface. Marking out the trunks *kidori* ("dividing the wood") in Japanese, always starts with a centre line (*shinzumi*) at the upper end (*sue-koguchi*). The master uses a plumb line and a carpenter's set square for this. The centre line is then transferred to the bottom end (*moto-koguchi*), which is usually about 10 cm greater in diameter, with a string (*mizu-ito*). If necessary a line can be made to run slightly eccentrically in this way, to avoid any damaged areas. Before the cross-sections are marked, the master carpenter applies more strings, to ensure that the timbers needed can be cut out of the trunk even if they are slightly crooked. At the top end the master carpenter does not just mark the cross-sections of the timbers, he also notes all the information needed for later allocation: the name of the building, a definition of the component, the component number and the number of the trunk. To do the marking, the master dips a bamboo strip (*sumi-sashi*), whose end has been cut into many small teeth to a depth of about 2 cm, into the ink-soaked cotton wool of his snap line. The marking starts with the largest cross-sections, and the smaller parts are cut out of the remainder.

The trunks are always marked out by the workshop's master carpenter. He is familiar with all the buildings and knows best what will be required of each individual part. As well as achieving the best possible yield, he also has to ensure that each component is cut from a trunk that is particularly suitable for the purpose. Thus for example beams are preferably cut from slightly crooked trunks, so that the round side of the timber can lie on top. Trunks with a high resin content are made into beams and purlins.

Quality classes are assigned to all building components in the timber list. The highest quality, used for making containers for storing the sacred objects and a few building components, is called *shihô-ake*, and the wood has to be immaculate on all four sides. Then come parts that must be without knots on two sides (*nihô-ake*). The quality used for minor and invisible parts is called *joko-bushi*, and here knots of up to 2 cm diameter are allowed. The timber lists also contain details about whether the cut cross-section will later be cut down into several sections or whether the parts will be fitted together to give a large cross-section.

End grain secured with wax and dogs; the incision to the core is well to be seen. These octagonal timbers will be used to make the posts to support the ridge purlin of the eastern treasure house.

Cutting the trunks with the band-saw.

Sheet copper for clad-
ding the post feet and
covering the timber
ends is prepared in the
sheet metal workshop.

Temporary wedges are
driven into the inci-
sion to the trunk core.

Cutting and storing

The trunks are cut up step by step by a large bandsaw, and turned many times in the
process. Here the daily output is between five and 15 trunks. After that they are loaded
on to a little railcar and pushed into one of the many storage sheds. Here the end-grain
is painted with a wax emulsion and dogs are hammered in to avoid cracking at the top
ends. The parts are then sorted by building and stacked carefully to dry.

Timbers from the trunk core (*shin-mochi*), as used for posts, beams and purlins, for
example, are cut through to the heart on one side (*sewari*) after trimming. Immediately
after that wedges are put in place and driven in at intervals of a few weeks. If the timber is intended for visible sections or to be in the immediate vicinity of the images, a
wedge-shaped strip is later fitted; this is glued on one side (*sewari wo umeru*) and then
cut flush. This elaborate treatment means that uncontrolled dry cracking is largely avoided.

Renewal stages

Up to 80 carpenters are engaged in the ten years before the renewal of the Inner and
Outer Shrine, some coming from distant regions. But there are only ten carpenters on
the permanent staff. Between the campaigns for renewing the two large shrines they
work in two groups, renewing or repairing the numerous minor shrines.

Each of the two groups has its own workshop in which about four minor shrines are
built per year. Here too the same care is taken as when cutting the timber. Cleanliness
is a key element of Shintoism, and this affects the work in the workshop. The carpenters all wear white protective clothing, a fabric cap with the emblem of the workshop
and usually cotton gloves as well, which help to avoid sweat marks on the untreated
cypress wood. When working on the central buildings they even start the working day
with a bath on the workshop site. The components are covered with white cotton mats
in the workshop.

After cutting, the round posts are first cut to a 16, 32 or 64 cornered cross-section,
according to size, before being hand planed round. For minor parts like the boards for
the fences and roofs the surface is now cleaned with a smoothing machine (*chôshiage-
kikai*); here the piece of timber is automatically guided past a fixed knife. Small quan-

This building site for one of the small minor shrines is completely enclosed during assembly, which takes about four days.

Building site for a minor shrine, fitting the gable decoration.

tities of screws are now also used in hidden places. For example, the laths used to cover the gaps in the board roof covering are placed over screw heads. Since the 61st Sengu the post feet have been completely clad in sheet copper, which slows rotting significantly and makes it easier to re-use the material.

The shrines are completely enclosed in a protective structure when they are being assembled. When the new shrine has been built, the image can be solemnly transferred to its new home. Special containers are made for moving and for storing the images. They are carved from a piece of wood, assembled from straight or bent boards. The old shrine is carefully dismantled soon after the ritual transfer. The valuable timbers are not disposed of, but re-used in a variety of ways, with the exception of the roof toppings and the feet of the posts.

Re-using timber

The two large gates on both sides of the nearby Uji Bridge over the Isuzu are made from the posts supporting the ridge purlin of the central shrine on the gable sides. At a diameter of 76 cm and a length of 11 m they are the largest building components. The posts are planed down for this purpose, and any damage is made good. After another 20 years the shrine passes them on to Sekimachi and Kuwane, where they are re-erected slightly smaller at the beginning of the old pilgrims' road to Ise after being reworked.

Post foot of a Torii at a minor shrine. The cut to the core has been filled with a strip of wood on the post, and a sheet copper bandage applied.

Minor shrine, detail of the pediment. The purlin is not placed directly on the post head. The load is distributed via the filling, to avoid shrinkage gaps in the board walls.

Some parts remain in Ise for work on minor shrines that are repaired after 20 years and now renewed only every 40 years. Timbers with smaller cross-sections are made in the shrine's own workshops into so-called god-shelves (*kami-danna*) of the kind found in many Japanese shops and homes.

Head of the Torii with wedged rail.

One particular building, the eastern treasure house (Higashi-Hôden) at the Outer Shrine, is dismantled and reassembled in the following year as Shinmei-sha on the island of Shinojima. All the parts are first smoothed by hand. The rotten feet of the supports, which were originally set into the ground, are cut off and placed on foundation stones. The reed roof is replaced by a sheet copper covering. This building is then moved again on the island after 20 years and used as Main Hall for the Hachioji-sha. But even then the timber is not thrown away; the large pieces are further converted and used to make 18 little minor shrines. So the wood serves for 80 years in all.

But the majority of the timber used in the Inner and Outer Shrine is passed to other shrines all over the country. A total of 169 shrines received timber from Ise after the last Sengu in 1993, some getting so much that they were able to build a complete new small shrine. For many shrines this is a unique opportunity to acquire high quality timber that they would otherwise scarcely be able to afford, but an additional factor is that the consecrated wood from Ise has a quality comparable with that of relics.

The pediment with
new roof surface and
ridge cover.

The bottom section
of the posts was
replaced.

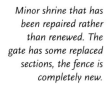
Minor shrine that has
been repaired rather
than renewed. The
gate has some replaced
sections, the fence is
completely new.

Workshop at the Great Shrine of Ise

Builders' lodge

The Ise workshops are a permanent builders' lodge with different trades and permanent building work, so they occupy an exceptional position. Otherwise there are affiliated workshops only at the Tôshôgu shrine at Nikkô and the Itsukushima-jinja near Hiroshima. In both cases periodic renewal of colour finishing is the main thrust of the work. The special charm of the Yamada workshops lies in the fact that all wood is dealt with by a single organization, from forestry via cutting to assembling the buildings. If one considers the relatively short time for which the buildings are used, the amount of time and material used seems disproportionate at first. It makes sense only against the religious background of the renewal of the housing for divine symbols. The buildings with their simple structure and ridge roofs look archaic; but as they are neither clad nor painted, heavy demands have to be met on the workmanship. The care with which the costly trunks are cut and processed is essential for the immaculate and pure quality of the shrines, which are doubly attractive against the background of the old woodland, which is left in its natural state.

Old and new shrine.

*Example of re-use –
Main Hall of the
Hachioji-sha on the
island of Shinojima,
the former eastern
treasure house of the
Outer Shrine.*

The tearoom is carpeted with tatami mats. On the right is the square alcove; an untrimmed pine was used for its post. To the left of it is the fireplace, and on the edge of the picture the host's entrance, behind which is the anteroom with the tea utensils.

Teahouses

Tea consumption came over from China in the 8th century at the latest, as part of Buddhist culture. Tea soon became the preferred drink for the priesthood and the nobility. At first tea was taken in living rooms or a screened-off area, it was not until the early 16th century that special rooms and little huts gained acceptance. Japan's early teahouses, which have come down to us only fragmentarily in descriptions and depictions, did not follow any building typology. But with the passage of time, schools emerged, and these raised certain tea-masters' preferences to a standard. The scope for design consequently became much more restricted, usually confining itself to putting conventional basic elements together.

From the 16th century tearooms and teahouses – probably as a reaction to the original grandeur – adopted an aesthetic ideal that is largely still valid today. The intention is to create a context for the tea ceremony that is as simple as possible, natural and conveys a sense of purity. The effort needed to achieve this kind of apparently random rural simplicity is extremely great, because everything is planned to make an effect, down to the minutest detail. The wide range of terminology used to describe ground plans, materials and surfaces is evidence of the high refinement of the tea cult.[11]

Although a small number of teahouses are now built in contemporary "translations" with modern materials and techniques, the traditional variants are still in the majority. But here traditional does not necessarily mean that the teahouses are always in classical gardens or that tearooms can be found only in traditional buildings. In very dense inner city conditions in particular they are now found on flat roofs or in apartment blocks. The Japanese are masters at ignoring things that Europeans might see as breaks or even charming contrasts; one only sees what one wants to see.[12]

Sono family teahouse, view from the south-west.

Site plan of the estate – the teahouse is situated in the inner courtyard of the three-wing house.

The most important centre for producing teahouses is the old capital, Kyoto. Here there are about ten workshops for architecture in the Sukiya style. As well as teahouses, Sukiya includes other architectural genres that have adopted teahouse design methods, especially homes, traditional inns and to an extent shop fittings as well. The Kyoto workshops do not just supply the local market: the city's prestige in almost all the traditional arts helps them to attract commissions from other parts of the country as well. One of the smaller workshops, and at the same time one of the few where planning and execution are carried out by the same firm, is Yamamoto Kôgyô's on the northern outskirts of the city, which built the teahouse taken as an example here.

Teahouse in central Kyoto

The estate in whose inner garden the teahouse was built in autumn 2000 is in the centre of Kyoto, about 600 m west of the town hall on a narrow side street running north-south. The main street is already dominated by modern office buildings, but in the side streets pre-war traditional wooden town houses are still in the majority. The client, who runs a planning office for bridge building, wanted to have a large apartment block built here, with a traditional tearoom on the top floor for his wife, a tea teacher. But things turned out differently: an old, intact well and two trees over 100 years old had survived in the garden of one of the three single-storey houses that had been pulled down. To preserve them, the idea emerged of building a traditional town house instead of the apartment block, in a U-shape around an inner garden with a teahouse.

The Yamamoto workshop

Yamamoto, whose workshop was recommended to the client, is now 70 years old. He learnt his craft as a carpenter in a small provincial firm that specialized in school and warehouse building. He returned to Kyoto after eight years and worked for another eight years under Master Usui, before starting out on his own at the age of 33. Since then he and his workshop have built a good 100 teahouses, of which only ten are in the city of Kyoto. Yamamoto is an entirely traditional craftsman. His own master taught him that the tea ceremony and the teahouse as its enclosing shell form a unit. This was perfected a good 400 years ago, and it is not to be changed arbitrarily. Yamamoto has passed on the old forms and hopes that his workshop will also continue it in the future, when he has handed the firm over to his successors.

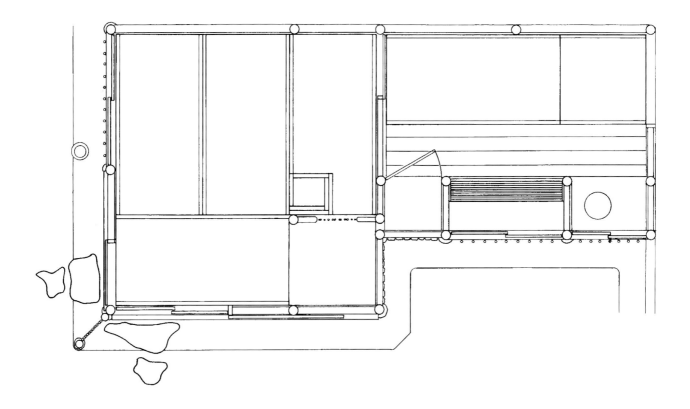

Concept

The client's wife spent years investigating historical and new teahouses. She was particularly interested in the Masudoko-no-seki teahouse, which was built in the 17th century in one of the many subsidiary temples of the Daitokoku-ji, a large Zen temple in north Kyoto. The name identifies the principal features of this teahouse, an alcove (*doko*), which is square, like the wooden beakers (*masu*) used for the rice wine. The historical model fixed the further development of the design process. The ground plan of the teahouse and also the design of the square alcove were adopted, but other elements vary according to local conditions and preferences. Examples of this are the position of the windows and the shape of the host's entrance, which is arched at the top here (*kato-guchi*). The teahouse is in the inner garden, and built on to the U-shaped

Ground plan of the tearoom with ante-chamber.

house. It consists of a tearoom (cha-shitsu) four and a half mats in size, and a room used for cleaning and storing the utensils (mizu-ya, literally "water-room") with sink, shelves and storage space behind little sliding doors.

Matsuban-Meiboku timber merchants; store with debarked and polished Japanese cedar.

Boards for the floor of alcove or suspended ceilings are stored upright and secured against warping with laths.

Sourcing and preparing the materials

Wood and bamboo for building a teahouse are selected with the utmost care.[13] Yamamoto buys them – when he cannot rely on his own stock – from Matsubun Meiboku, the timber merchant in Sembon-dori, which has been a street with numerous *mei-boku-ya*, literally "shop for famous wood", since the twenties. These are timber merchants who have specialized in choice building timber of the kind used in visible areas in high-calibre structures. Matsubun Meiboku is known for his excellent selection of timber that is essential for teahouses and related architecture, i.e. polished round timbers (*migaki maruta*) for the structural frame and posts (*toko-bashira*) and planks (*toko-ita*) for the alcove. He is able to cover part of his range from three large parcels that he runs in Kitayama, the mountains immediately north of Kyoto, where the famous Japanese cedars grow.

These cedars, which are used for the posts, head tie beams, purlins and rafters, are felled between the 10th and the 20th of September, when the trees have practically stopped growing. The bark is provisionally removed with scrapers and the trunks then remain lying at an angle for a month, with their crowns still attached, so that the water can evaporate via the needles. About ten trunks are leaned and tied against a tree, look-ing like the structure of a round tent. In late October the trunks are cut to 3 m lengths, transported, and then the rest of the bark is completely removed either by water pres-sure or by hand, according to the quality of the timber. After that they are left in the open air for an intermediate one-week drying phase, before they are put back into a bath of water so that the last vestiges of bark can be removed with a special knife. The trunks are then rubbed down with sand, which gives them the requisite silky sheen.

Specially debarked cedars with a characteristically uneven surface (*shibori maruta*) are held in particularly high esteem for the alcove posts. Originally this was a mutation, but the effect has been produced artificially for a good 80 years now by binding bamboo or plastic strips around the trunks of 25 to 30 year old cedars two years before they are felled. This prevents the trunk from growing outwards, the strips cut into it and leave a fluted surface. The strips are removed again in spring, and the tree is left to itself until it is felled in autumn. Matsubun Meiboku has a selection of nearly 3000 such posts for alcoves, costing between € 1000 and € 5000. The preferred timber for the head tie beams (*mawari-buchi*) is unstripped Kobushi wood, as its bark is considered particularly attractive.

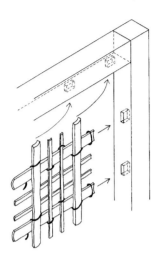

Manufacture in the workshop

When the material has been acquired, all the parts are marked out and all joints cut in the workshop. For a small building like this one the feet of the posts are also immediately adapted to the uneven surface of the foundation stones. To do this, the carpenter measures the surface with an implement like dividers, then transfers the profile to the post foot, which he then cuts down with a gouge. To make fine adjustments, the foundation stone is rubbed with chalk. The contact points then show on the post foot so that material can be removed until a perfect fit is achieved. The joints are already cut in the workshop. They take a great deal of laborious work as the irregularities of the round timbers used have to be taken into account. The parts are covered with paper (using a special glue made of sea-grass) from the time when they are prepared until the building is

The broad bamboo strips for the support structure have already been put in place.

assembled; this protects them from damage and from sweat marks. All the parts have to be handled with great care anyway, as the wood is not painted.

After completion of the bamboo support structure; the right-hand wall has already had clay applied to it on one side.

Assembling, filling in, covering

Once the foundation stones are in place, the structural framework can be assembled in a few days. Then the framework bays are filled in.[14] First, bamboo strips a good 2 cm wide are placed at intervals of about 30 cm, and these are tenoned into posts, sill, tie beam and head tie beam. Then a close-meshed lattice of finer bamboo strips is tied on with rice-straw string. Unlike European practice, the supports for clay filling are not woven, but tied one on top of the other in layers. Now the time has come to create a kind of windows in the support structure (*shitaji-mado*), by leaving it unrendered at certain points. Originally they were a solution used in farmhouses for making an opening for light and ventilation with the least possible effort. In teahouses these windows do not have a split bamboo support structure, but instead thin bamboo tubes are carefully tied into a fine grid with withies.

Erecting the structural framework. All the visible structural parts have paper stuck on them.

Reed grid for an unrendered window aperture.

Constructing the lower ceiling shell in bamboo and shingles.

View from the south-west during building.

The support structure in the bays is rendered twice on both sides with clay that has been mixed with rice straw and soaked for several months. For this teahouse the customary third application of a specially fine clay rendering was omitted, as an unduly perfect surface would tend to spoil the rustic ambience. The teahouse was roofed with pantiles and covered with sheet copper on the south gable side. Here it deviates from its historical models for reasons of cost and so that it will last for longer. Formerly it would have been covered with bark or timber shingles.

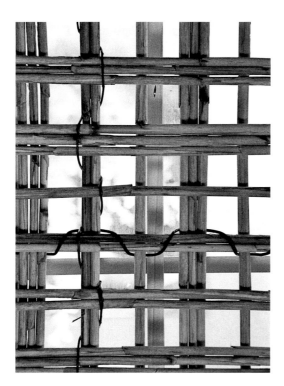

Detail by the unrendered window aperture on the west wall.

View from below of the ceiling in bamboo, reed and cedar shingles.

Woven split cedar strips at the point where the guests wait on a bench.

Ridge detail; the ridge purlin and rafters were made of cedar debarked and then polished.

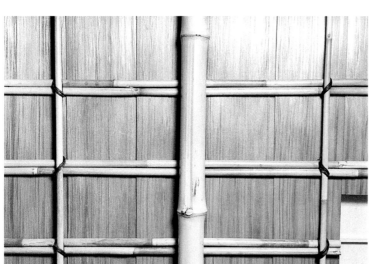

Interior finish

The ceiling of the teahouse, which is divided into three areas, required a great deal of work. Seven bamboo poles to support a grass mat (*makoma*) were laid east-west in the north section, with its area of three mats. Originally bamboo from the roofs of old farmhouses was used for this purpose. Decades of smoke from the open hearth had coloured them dark brown (*susu-dake*). This material was particularly highly esteemed, as the tea-masters felt that it created a venerable and worthy atmosphere. As there are scarcely any farmhouses with open fireplaces left today, the effect has to be achieved by smoking.

The ceiling above the alcove was set about 30 cm higher. There is no ceiling above the south-western mat, and so the roof diagonal can be seen from below (*kaegomi tenjo*). Strong bamboo tubes with the knots deliberately left on them (*metsuki-no-shiratake*) support two layers of bamboo tubes laid crosswise (*medake*), and thin, split cedar boards (*hegi-ita*) are laid above these. The difference in height between the part of the ceiling covered with the grass mat and the part where the ceiling follows the roof diagonal is used somewhat unexpectedly for the air-conditioning slit, as the teahouse is attached to this system.

Mizuya – anteroom with sink and shelf for the tea utensils.

Tearoom – in the foreground on the left is the fireplace with suspended kettle, there is an unrendered window aperture in the protruding wall, and the small, square opening for guests is in the rear wall.

The lower parts of the clay-rendered walls – with the exception of the alcove – were covered with paper (*koshibari no kami*). This protects the clay rendering, but at the same time is an important design element for enlivening the walls. The vulnerable edges of the clay rendering are also reinforced with paper at the host's entrance. Here old paper with handwriting on it (*hogoshi*) is applied; the front is always stuck down, however, so that the writing can scarcely be deciphered by the guests, and does not distract their attention. Otherwise dark blue paper was used (*minato-gami*). The two kinds of paper are applied to different heights, to make the space more interesting.

Low square entrance for guests. Originally part of a so-called rain door was re-used for this purpose.

With the exception of the small, square alcove in the corner, the floor of the teahouse is covered with rice straw mats. Here the tatami still have a straw core (*waradoko*), which has now become quite a rarity. One of the four mats has a small piece cut out of it for the fireplace (*ro*).

The host's entrance has a sliding door made of a thin wooden grid covered with paper on both sides (*taiko-busuma*, *taiko* is a drum, *fusama* means sliding doors with paper on both sides). Guests come to the teahouse through the inner garden, then crawl through a low, square aperture (*nigiri-guchi*) on the east side that is closed by a small sliding door. The grid on this door, which is clad with thin boards on the outside, is divided up remarkably unevenly. Originally part of a large sliding door was used for this purpose, of the kind used to protect house verandas from the rain (*amado*, rain door). So something that started as a way of using an old building component was stylized and is therefore elaborately manufactured today.

Well, lantern and one of the two 100-year-old trees from the inner courtyard of the previous building.

The alcove

The post for the alcove (*toko-bashira*), which stands in the room, is of central significance. Here a carefully selected, unstripped pine trunk (*akamatsu no kawatsuki*) was used. On the two sides facing the middle of the room the trunk was slightly trimmed to a height of about 40 cm. This creates a lancet-shaped section at the foot of the post. As the shape is reminiscent of young bamboo, this treatment is also called "bamboo shoot" (*take no ko*). The alcove post has an incision (*sewari*) on its rear side to prevent the trunk cracking because of dryness. The actual alcove, whose area is exactly half a tatami mat, consists of a solid pine floorboard about 30 mm thick, secured against warping by two battens with sliding dovetails on the bottom side. This floorboard, about 90 cm wide, is not glued, and particular importance is attached to attractive grain when choosing it. A thin clay screen is set between the alcove post and the east wall. It has a window created by a gap in the clay rendering in its central section.

Three nails are driven into the alcove to alter its appearance according to the season and the guests, a "flower nail" (*hana-kugi*) for hanging small baskets or vases of flowers on the post, a "middle nail" (*naka-kugi*) placed centrally in the rear wall and a "bamboo nail" (*take-kugi*) in the rear wall for large picture scrolls.

Exterior view.

A path with stepping-stones and a gate lead to the teahouse.

Sublime rusticality

This teahouse presents astonishing variety within the most restricted space. This applies first of all to the material. In the framework, polished cedar trunks were combined with head tie beams in unstripped Kobushi wood (north wall) and an untrimmed pine trunk for the alcove. We also find variations in material and form in the ceiling and the complex roof. The ceiling is divided into three areas created with different materials and techniques, and the roof is a combination of two staggered gable roofs and shed roofs on the garden side. The sliding doors and sliding windows are also quite different. The host's door is covered with paper on both sides, and the guests' entrance imitates a rain door; only the two doors on the south side, which are only opened to provide light and a view of the garden, are *shôji*, in other words sliding doors covered with transparent paper on one side.

Originally rural elements have been sublimated with a great deal of effort and craftsmanly perfection for the structure. This architecture is committed to being unpretentious, but achieves its simple overall image only with the greatest refinement. So the price of this product is not surprising; the little four and a half mat teahouse cost a good € 300,000, thus more than a third of the price of the town house around it, with a living area of about 100 mats.

House building

Japan's earliest dwellings were earth-pit houses (*tate-ana-jûkyô*), which are known only through excavations. A simple post framework was set up in a pit about one metre deep and rafters laid against it, which made the houses look like tents. Buildings with raised floors appeared in the second century at the latest. They were used for storage at first, then soon after as houses for the nobility. Down to the 6th century, the posts were fixed rigidly into the ground. With the adoption of Buddhism and continental forms of architecture, however, this construction was gradually replaced by a system of columns standing on foundation stones or sills. They became much more durable by the use of raised floors, posts standing on foundation stones and therefore no longer in contact with the ground, and also the widely protruding roofs protecting the structure against rain and sunlight. Although this building method was soon also used for temple and palace architecture, it took more than a thousand years before it was generally applied even for simple structures like farmhouses.[15]

The structural frameworks for houses, which usually have a single storey, and never cellars, consist of posts linked on several levels by penetrating and non-penetrating tie-beams. The posts, which can be placed on individual foundations or on a ring of sill beams, support the head tie-beams, above which are the roof beams. The roof structure above this is usually a forest of closely placed struts, which support the purlins. As there is so much structural timber in the roof space, it is not possible to use it for anything else. Among the few exceptions are the steep saddle roofs in the Shirakawa region, where the attic is used for breeding silkworms; this is only possible because the quite different gabled roof structure with its large open space is used.

The characteristic open structure of a Japanese house with few permanently defined areas, and many that can be closed off or opened up, reacts to the hot and humid

Exterior view – the Yoshida house faces east.

climate, as it makes adequate air circulation possible. From the 13th century onwards, sliding doors and windows become increasingly numerous, and appear in many variants. If necessary, they can be pushed back or removed completely, which is why the ground plan can be varied so easily. In this way, several small rooms can be made into a bigger one without a major effort.

The floor is laid with thick rice-straw mats, which are covered with woven rushes and, at the edges, with a strip of fabric on the two long sides. At first thick mats were used as seating on the wooden floors at court, for example, and could be moved around at will. Later such mats were laid on the periphery of large halls, then from the 15th cen-

The gable is constructed in the same way as the roof: a large number of short struts are placed very close together and joined by penetrating tie-beams.

The windows on the north side of the house have small protective canopies. The outside walls are clad in weather-boarding to about half their height.

tury the whole floor was covered with them. Standardized mats, exactly twice as long as they are wide, which could then be used as a plan module for measuring the ground area, appeared in the late 16th century.

Furniture is reduced to a minimum in the house, there will be only low tables and perhaps a few cushions to sit on. This is possible because there is no seating as such, and no bedsteads; there are also a large number of wall cupboards, without which practically no room is complete. Other storage space was usually offered by one or more small fireproof storehouses for provisions and valuables. The reception room has a particular interior, with a picture alcove and sometimes also an alcove with staggered shelves and small wall cupboards, and also a reading and writing alcove.

Even after 1868, when the country was opened up and modernized along Western lines, the traditional Japanese house showed a remarkable degree of ability to persist. Public buildings, factories or transport facilities were quick to follow Western models, but homes continued to be traditional. Only few members of the upper class built themselves Western-style homes, though quite a lot of people added little extensions to their traditional houses, with a room furnished in the Western manner. It was only after the Second World War that this ratio was to change. Today most homes are furnished in the Western style, with only a single Japanese room. Practically no traditional houses are built in big cities, but there are rural areas – albeit only a few – where these traditions persist. One area in which prosperous farmers in particular still build houses like this today, and where people have for centuries set great store by houses that are as impressive as they are durable, is Hokuriku on the Sea of Japan. It is not unusual for these buildings, with their size, their complex roofs and lavishly decorated detailing to have something of the nouveau riche about them.

The Kiya workshop

A carpentry shop that specializes in building traditional houses, designed with a pleasing austerity, is Kiya in Fukkô, 40 km south of Takaoka. The family name also signals the programme for this workshop: Kiya is written with the two characters for wood (*ki*) and house (*ya*). With the exception of the nobility, top military personages and rich citizens, most Japanese families did not have surnames until the late 19th century. It may be that the family chose this name deliberately; it is certain that they have practised the profession of *ya-daiku*, carpentry for house-building, since the mid 19th century.

The building sites are within a radius of a good 10 km in the west of the Toyama Prefecture. Their commissions come mainly from farmers. Until the Second World War it was the rule that timber for building a house should be felled in one's own forest. Even today a good third of the clients have their own woods, and use this resource to build their homes.

The Yoshida house

In 1999, Kiya was commissioned to build a new house for the Yoshida family, who had become prosperous through making organic fertilizer. The house is now occupied by eight people from three generations, and with its almost 300 m² of living space it is unusually spacious, even in this region, that is known for its large houses.

The house faces east and has an L-shaped plan. It consists of a two-storey main building (*mo-ya*, literally mother-house), around which are single-storey extensions on all four sides (*ge-ya*, literally lower house) with shed roofs. A single-storey wing on the north-east side accommodates bathroom and kitchen. All roofs have shallow slopes and are covered in a type of interlocking pantile. A striking feature, typical of the region, is

Views from east and north.

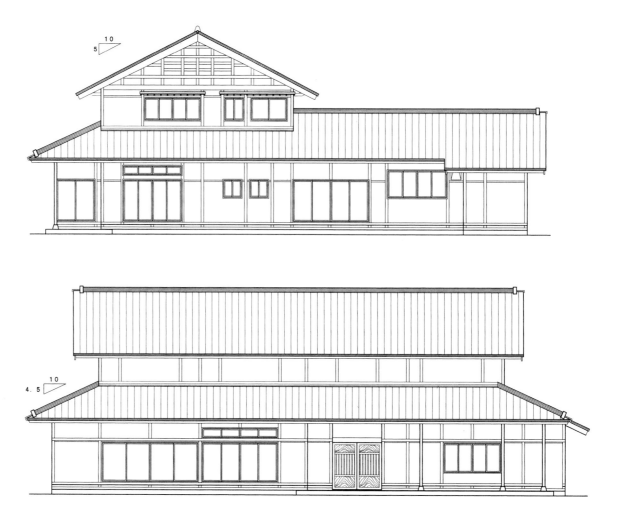

the broad pediment on the main roof, in which closely-spaced struts are linked with several penetrating tie-beams, thus forming a close-meshed framework. The veranda windows and doors have little canopies. The weather-boarding reaching up to a height of about 1.2 m on the ground floor is also intended as a protection against the elements.

Plan

The guest entrance is on the east side of the house. The great hall (*genkan*) is beyond two wide sliding doors; the first third of it has a floor at ground level. Here one takes one's shoes off and a step leads to the rear area of the room (*shiki-dai*), which has a wooden floor. The fact that the living area is on a higher level is reflected by the way guests are greeted in Japan: they are not asked to come in, but to come up. Immediately adjacent to the hall is the 15-mat living-room (*hiro-ma*) with an open fireplace and shelf for the family gods (*kami-dana*). Further south are two ten-mat rooms, a front living-room (*kuchi-no-zashiki*) and a back living-room (*oku-no-zashiki*). Like the large living-room, the front living-room also has a veranda in front of it, giving the room greater depth and a view of the garden, and thus mediating between interior and exterior space. The rear ten-mat room accommodates the picture alcove (*tokonoma*) and a writing alcove (*shoin*), and also a wall cupboard containing the family's house altar. Almost every room has a wall cupboard one mat deep, and thus disposes of a great deal of storage space.

Ground plan.

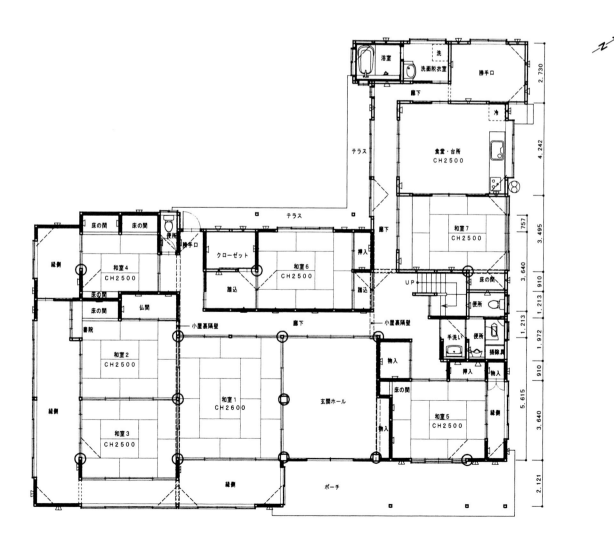

Structural framework

The house stands on a ring of sill beams set on low concrete foundations (*nuno-no-kiso*). Until the post-war period, all posts had individual foundations in the form of large, undressed field-stones found along the nearby riverside. The house has a large frame structure at its core, an architectural feature of this region (*naka-waku-zukuri*, inner-frame building method). Massive posts, rising through two storeys and placed wide apart, are joined above the sliding doors by a lintel a good 60 cm high, and above that by several penetrating tie-beams. This core frame accommodates the two central rooms in the house, the hall (*genkan*) and the adjacent "Great Room" (*hiroma*) with an open fireplace. Originally both rooms were of double height, but in the Yoshida house only the hall has retained this feature, while there is another room in the upper storey above the large living-room. Only Keyaki wood is used for the core framework. Some trunks with crooked shafts were used for the beams, squared up on two sides only. All the trunks were smoothed and then coloured, a particular feature of the region, in which the lacquer metropolis Wajima is located. Each one is painted once with iron oxide (*bengara*) and fermented Kaki juice (*kaki-shibu*), before being given seven coats of Urushi varnish. This gives them a brownish-red hue, and a deeply glowing surface.

Entrance with a large stone step and two wide sliding doors.

Two-storied entrance hall.

Decoration for the Girls' Festival.

Ten mat room with picture alcove and wall cupboard for the Buddhist house altar.

Building process

Kiya uses different kinds of wood for the framework of the house. The core frame is in Keyaki, and cypress is used for the short posts in the one-storey extensions and also for the ring of sill beams. Japanese cedar is preferred for the door sills, lintels, the rails and also the purlins in this region, in this case from the client's woods. All the timber is mechanically planed and cut to size in the workshop after preliminary cutting in the local sawmill. Kiya still marks the joints with snap line and Indian ink. The preparation of the sections, cutting and test assembly of the joints took almost 1500 man-days, but the framework of the house was actually built by only seven men in a single week. This ratio of elaborate preparation and rapid assembly applies to all building jobs in traditional Japanese timber building. Once the supporting framework is in place, the roof is covered (*yane-kôji*). The rafters are shuttered over and then covered with widely spaced wooden shingles. Next a thin layer of clay is applied to take the tiles, which are a kind of pantile. In the early days only temples, palaces and important public buildings had an

77

Assembling part of
the inner frame in
the workshop.

*The wood has paper
fixed to it to protect
the surfaces finished in
Urushi varnish during
assembly.*

elaborate and originally very expensive hard, tiled roof like this. In the 17th century they started to be used for urban houses, and since the early 20th century for farmhouses as well.

Next come the renderers (*sakan-kôji*). In our example the solid panels in the framework were closed in traditional clay. To do this, a split bamboo support is inserted. Its horizontal and vertical strips are not woven but tied together with rice-straw string. The clay is mixed with chopped rice straw, soaked for a long time and applied in three stages. A start is made by applying coarse rendering on both sides (*ara-kabe*), after drying for at least two weeks this is followed by intermediate rendering (*naka-nuri*), before the concluding fine rendering (*shiage-nuri*) is applied after another three weeks. Kiya allows about 200 man-days for the rendering work.

When the rendering has dried sufficiently, the carpenters come back and work on the weather-boarding and small canopies. Finally work begins on the interior, the floorboards are laid and the ceiling hung. Keyaki left over when cutting the sections for the core framework was used for the floorboards in this house.

*The frame of the
house was put up in
a very few days.*

The many sliding doors in traditional Japanese houses are always made by joiners. In this house they were mostly made by hand, which is now a great exception. On rare occasions veranda doors are made of wood, as was the case here; since 1965 aluminium frames have taken over. Special care was taken with the wide doors for the large room and the hall. They are called *obi-do* after the 12-13 cm wide central rail that divides the door into an upper and a lower panel, and is reminiscent in dimensions and position of the broad Obi belt that holds the kimono. The panels are about 10 mm thick and 130 cm wide and cut out of a single unglued board (*ichi-mai-ita*). The four large Keyaki panels alone cost about €15,000. The desire to avoid gluing even such large planks and boards comes partly from a time when only bone and rice glue were available and offered scant guarantee of durability. But it was above all creative ambition, the urge to show the spectacular grain pattern of the wood like a picture, uninterrupted by a single glued joint, that led to such effort and expenditure.

Master carpenter Kiya.

There are some particular features that distinguish the house described here from its predecessors. Older houses in the region used wood more frugally, long parts like the rafters were put together from several pieces in order to achieve the required length, and the number of round timbers was greater. Bath and toilet used to be sited in the north-east corner of the house, while here the two toilets have moved closer to the living quarters. A striking feature is the relatively high proportion of halls with wooden floors, as a consequence of the large area occupied by the house. In older houses the rooms were accessed mainly via the side rooms or the veranda. Yet despite these differences, the Yoshida house is firmly rooted in the local building tradition, which has been broken off in most regions of Japan.

The complete house framework before the wattle-and-daub panels are filled in.

Architectural models

When Buddhism was adopted in the mid 6th century, continental building forms that were considerably more complicated than the architecture built up to that point gained a foothold in Japan. Buildings now reached a gigantic dimension, and it was not just the number of components that was increasing, the constructions were more complex as well. This leap can be seen particularly clearly in roofs, where curves with a clay tile covering replaced simple gable roofs covered with reeds or bark. Even today, building such complex structures demands intimate knowledge of construction from builders, especially as the parts of the supporting framework are all prefabricated, so that the building can be assembled very quickly.

Preparing such large building projects in temple and palace complexes will have triggered the making of the first models. The earliest example is the temple of Hôkô-ji in Akakiban, about 25 km south-west of Nara, completed in 596 (not extant today). The craftsmen from the Korean kingdom of Kudara, under whose direction Japan's first Buddhist temple complex was built, are said to have brought a model of the main hall with them, according to contemporary sources. And several tiny building components like bearing blocks and bracket arms from cantilevering timberwork said to come from models were found at excavations on the Nara palace site (*Heijô-kyô*; capital from 710-

784).[16] Two complete models dating from the mid 8th century have also survived, the "little pagodas" from the Kairyûô-ji and Gangô-ji temples in Nara. Both are five-storey pagodas, made on a scale of 1:10. For the 4.1 m high Kairyûô-ji pagoda the construction of the second to fifth storeys was simplified to the extent that the walls are carved from a single piece of wood in each case (*hako-zukuri*, box construction method). The 5.5 m high Gangô-ji pagoda is different; here the entire structural inner life is repro-

duced in detail. So the function of the models lies in realistic reproduction not just of the building's envelope, but above all of its construction.

The model of the well-known Shôsôin storehouse is assembled on the top floor of the workshop.

Similar construction models have survived from several epochs, including for example another small five-storey pagoda in the possession of the Kyôôgokoku-ji temple in Kyoto (14th century) or the keep of Odahara castle(16th century). That their function was as a plan substitute or to complement a plan is all the more plausible as detailed elevations of buildings are not recorded until the 16th century.[17]

The 1897 act protecting old shrines and temples made conserving the architectural heritage a public duty in Japan, and from then onwards the condition of the buildings before and after restoration was recorded in detailed plans. From that time, occasional models were made of complexes, individual buildings or details, to show the original condition (one example here is the 1909 model of the Tôdai-ji temple; others are models of sections from the Nandai-mon and Tegai-mon gates at the same temple made from 1927 to 1930). Models were also built to examine a return to the original form in the course of a restoration. Finally the possibility was recognized of making models of important buildings as records and as teaching aids for the study of architectural history. Like their oldest predecessors they are usually made on a scale of 1:10. Since 1965 the Agency for Cultural Affairs, the highest authority for the protection of cultural assets, has had a special budget for making architectural models. Most of the models made since then come from the Wada workshop in Akashi, whose work is presented here.

Trimming the log
timbers with a
smoothing plane.

The end grain is also
trimmed with the
plane.

Wada workshop in Akashi

Yasuhiro Wada, born in Kobe in 1932, and his brother Yûkô, who is four years younger, come from an old family of carpenters. Both trained with an uncle for five years after the war before entering the service of another relative who built shrines and temples. From 1956 to 1964 the two brothers helped to restore two medieval halls, the Monju-dô hall at the Nyoi-ji temple and the main hall of the Taisan-ji temple. They must have drawn attention to themselves with their particularly good and precise work, in any case the site manager asked them to build the model of the corridor, whose existence had been discovered by excavation, of the former palace complex at Nara. The two Wadas made models on the best-known buildings from 1965 onwards. In the meantime 30 such copies have been made in their workshops, mostly commissioned by the Office of Cultural Affairs. They include such well-known buildings as the Great South Gate at the Tôdai-ji, the Jôdo Hall at the Jôdo-ji and the Silver Pavilion in Kyoto.

Building a model

For trimming small
cross section parts, thin
strips are mounted on
the edges of the plane
sole. A piece of hard-
wood is inset in front of
the mouth; under it are
two springs that put
pressure on the light
work-pieces.

The starting-point is a set of plans drawn up as part of a comprehensive restoration of the particular building and kept in the archives of the Agency for Cultural Affairs. To complement these documents, the craftsmen usually travel to the original for a few days, to clarify details.

The models are usually made on a scale of 1:10. Wada uses Kiso-Hinoki for them, very close-grained cypress wood from the Nagano Prefecture. He chooses a suitable trunk from the timber merchant and supervises the cutting at the sawmill. The timber is then kiln-dried at the mill.

As for a full-size building, the parts are all prefabricated for the model as well. Rough cutting is done by machine. But the parts are planed to their final dimensions by hand, with a home-made plane. On both edges of the sole it has a strip mounted on screw-heads at a height corresponding with the final dimensions of the finished piece. A hard-wood slat about 9 mm wide is placed in front of the mouth, and underneath this is a double steel spring. Thus pressure is placed on the thin, light piece of wood while it is being planed. Planes like this – the model-makers learned how to construct it from a joiner specializing in sliding doors – are jokingly called *hikôki-ganna*, literally aeroplane planes. The parts are not marked out with ink but with a little knife. They are then cut to length with a small crosscut saw with a very fine blade, and left a little larger than the final dimensions, so that the top end can be trimmed with the plane or a small knife (*ko-gatana*).

The greatest difference from the building site is probably the sequence. When building a large hall, the structural framework and the roof frame are put up first, and the build-

ing is then covered. Only then the bays between the structural framing are closed and the interior finishing is done. But the model-maker has to work strictly from bottom to top. Often the parts have to be coloured before they are assembled, because the interior will be inaccessible later.

The model tries to be as close to the original as possible. As has been said, the aim is not just to copy the envelope but to follow the original construction and use the original materials as much as possible. The relatively large scale makes it feasible to construct even the complicated joints. For this purpose, Wada has cut some pins from tatami workshops down to particularly small chisels. If the original has a roof covered with organic materials, the model is also covered with strips of bark, wooden shingles or reeds; only clay tile roofing is substituted in the model with small carved tiles made of wood.

With the exception of pagodas and small gates, Wada's models are "divided models" as a rule. They are in two halves, with a thin joint running down the middle. If needed, the two halves can be taken apart to give a view of the section of the building with its complex structural internal life. The only structural differences affect beams and rails: in the original they are usually lengthened using several pieces and elaborate joints, as very long building components are difficult to come by and assemble. But a divided model would be unstable if it were made up of too many jointed parts; for this reason these components are made in one piece.

The work is roughly divided up as follows between the three model-builders – one of the brother's sons has been working in the workshop as well since 1996: the uncle is particularly good at making the structural parts, the father does the carving, usually working with magnifying spectacles, and the son is responsible for the finishing, sometimes in colour. One particularly charming feature of their work is that they are masters of more than one trade; a model-maker does not just need a carpenter's skills, he also needs the knowledge of a roofer, painter and smith. Precision is the supreme requirement for all the work, as unlike the original it is possible to see practically all the parts of the building. There is nothing to be hidden, you cannot run away, as Japanese craftsmen say (*nige ga nai*).

Protecting techniques

Half-finished model of the Daizuhô-den hall.

Since it was amended in 1975, the Japanese law on the protection of cultural assets has also included the possibility of protecting and promoting techniques that are essential for conserving such assets. Making architectural models has been recognized as such a technique since 1994. Wada is probably the best-known exponent, and so the education minister chose him to be representative of this technique, and since then he has received an annual grant of about 1.1 million yen. Wada sees the public esteem in which model-making is held and his personal recognition as a driving force behind the technique as a commendation of all model-makers. But paradoxically he has received less commissions since the award, because prefectures and local authorities have had to cut or even cancel their model-making budgets because of excessive debt.

Completed hall.

The model can be taken apart into two halves to give a clear view of its complex inner construction.

Wooden shingle and bark roofs

Roofs are much more important in Japanese architecture than they are in Europe, for example. They not only offer protection from the weather, their relatively great height also makes them crucial to the overall impression made by the building, and they often indicate social status and public authority. Temples, shrines and palaces have particularly elaborate roofs. Here the shapes are complex, as in the typical hip-and-gable roofs, for example, and they are elegantly curved. The roof diagonal is usually kept slightly concave here, and the eaves rise somewhat towards the corners of the buildings. Ridge and hips are often raised and decorated in a variety of ways.

There are a large number of materials and techniques available for covering a roof: reeds, straw, boards, shingles, bark, sheet copper, led and tiles. Roofs covered with wooden or bark shingles are rarely found in Japan today, on temples, Shinto shrines, teahouses and old palaces.[18]

Roofs covered with planks have existed from the 6th century at the latest. But today we find roofs covered with boards (*ita-buki*) only on a very few buildings like the subsidiary shrines in Ise and in the extensions to the lower storey of the main hall and pagoda at Hôryû-ji near Nara. There are records of bark roof coverings from the 8th century. In the Heian period (794-1185), as contacts with the continent were weakening and architecture was growing away from its models, the villas of the nobility were often covered with bark. None of these buildings has survived, but they are depicted in minute detail on the picture scrolls of the period.

In the 10th century, the roof structure of religious buildings changed radically, with a single-shell being replaced by a double-shell structure. The lower shell with rafters set very close together was essentially decorative, while the roof covering was carried by the so-called concealed rafters of the upper shell. This change gave the master-builders new creative freedom, as the angle of the two roof shells can be fixed independently of each other, and the eaves line raised somewhat. This produces the curve that is so typical of Japanese roofs (*sori* or *mukuri*; well-known early buildings with double-shell roof structures are the Hôryû-ji Teaching Hall and the Phoenix Pavilion of the Byôdô-in near Kyoto). Diagonal cantilever beams (*hanegi*) were soon arranged between the two shells of the roof structure, transferring the load from the widely protruding roofs to the posts. This means that the roof protrusions are now largely immune to the notorious threat of settlement. The consequence is, however, that a gap occurs at the eaves, which has to be closed. In the case of bark and shingle roofs, shingles have to be layered one on top of the other at the edges of the roof.

Wooden shingle roofs

Roof coverings made of wooden shingles are found above all in three types of buildings. The hip or hip-and-gable roof of a temple hall is frequently covered with them. From the 16th century they were the preferred material for the villas and teahouses of the nobility, the best-known example being the Katsura villa in Kyoto. Until they were displaced by the tile roofs that became compulsory after the great town fires, town houses sometimes had shingle roofs as well. In some regions, along the Kiso road, for example, which were for a long time centres of wooden shingle production, they were partly also used for homes up to the mid 20th century because they were so readily available.

A distinction is made between three variants, according to thickness. A roof covering with shingles 10-30 mm thick is called *tochi-buki*, and roofs of this kind are found only on about 30 listed buildings, many of them in the north of the main island of Honshû. For especially high-calibre buildings like for example the palace of the abdicated Tenno (*Sento-gosho*), builders liked to use shingles 4-6 mm thick (*tokusa-buki*). But most frequently the shingles are thin, about 3 mm thick, and called *kokera-buki*. And additionally wooden shingles – though with rows that are further apart and less elaborately worked – are also used as the lower covering for tiled roofs. Called *doyo-buki* (also *tonton-buki* or *doi-buki*), they are first pinned on to the support structure before a layer of clay is added and the roof tiles are laid in this. At first only important public buildings and temples had such clay tile roofing, but from the 16th century town houses were roofed with tiles as well.

The Sugimoto roofing firm

One of the oldest roofing businesses is Sugimoto in Kyoto; the family is said to have been in this profession for 700 years. To continue this line, Sugimoto Sôhei (born in 1932) was adopted by his childless uncle when he took over the workshop in 1980. Here shingles and bark are used for roofing, but wooden shingle roofs are the mainstay, and many of them cover the neighbouring temples. Some little wooden panels hang in the hall of the parent house, listing the names of the most important clients: Daitoku-ji, Myoshin-ji, Kitayama-Kinkaku-ji, Ninna-ji, all very well-known temples in the north of the city. Until a few years ago, many commissions for lower roof coverings came for new tile-roofed homes. But these numbers have been decreasing for years, and collapsed after the great earthquake of 1995 in Kobe, when traditional tile roofs fell into disrepute because they were so heavy. Today more than 90 % of Sugimoto's commissions come from shrines and temples in the Kyoto area. The rest are teahouses and occasional reconstructions of historical buildings.

Making wooden shingles

The timber used for shingles was easy to split, durable and happened to be available locally. Thus for example the shingles in the Izumo region were made mainly of chestnut, and of Japanese cedar in the northern part of the main island of Honshû. Because the timber stocks have shrunk considerably and prices of Japanese cedar have increased, Sawara Cypress is used for shingles now. This conifer, very similar to the cypress, is easy to split and smells pleasant. Only the small roofs of teahouses are still covered with Japanese cedar shingles.

Unlike Europe, where wooden shingles were already being sawn in the 19th century, they are still split in Japan. The unseparated fibres are needed for durability. Water quickly runs off along the fibre and the roof covering dries quickly thanks to the thin layer of air between the shingles; water is not drawn in by capillary action, as is the case with sawn shingles.

Sugimoto buys his wood by the trunk from a Kyoto timber merchant. A good 80 % are Sawara trunks from the Nagano Prefecture. They are processed six to twelve months after felling. Outside the workshop is a Sawara trunk about 6 m long and just under 1 m

Thick logs are split off from the segments. Their thickness is measured with a bamboo gauge.

The logs are placed on edge and split down the middle.

in diameter. After the master has marked the length with chalk crayon, discs 60 cm thick (*tama-giri*) are cut with a chain saw. Until 1980 this was still done with a two-man hand-saw (*ryô-tebiki-noko*). The disc of wood, which corresponds to the length or twice the length of the shingles, is first halved with wedges and a sledgehammer, and then further divided into six or eight segments. This is called *mikan-wari*, as the disc is taken to pieces like a tangerine (*mikan* is the Japanese mandarin, *waru* means splitting).

A section of a large trunk is set into the floor, protruding by a few centimetres and serving as a base for splitting. Then a splitting knife and a large rubber hammer are used to split off the sapwood and heart, which are later burned; the sapwood is softer and less durable because of the substances it contains, and the heartwood is hard to split because it has a large number of little knots (*shim-bushi*, heart knots). A simple gauge made of a strip of split bamboo helps in placing the splitting knife correctly so that sections can be split off in a thickness of eight or 16 shingles (24 and 48 mm respectively); this working step is called *bun-dori*. The master trims the logs on all four sides with a circular saw; in earlier times they used to be trimmed with a two-handled drawing knife (*sen-bôchô*).

Sixteen shingles are now cut from every log; this process is called *ko-wari*, or fine splitting. A thick beam with a deep cut in it is nailed to the top of the trunk section in the floor, to form a base for the splitting (*wari-dai*). The craftsman places the log upright alongside the beam and holds it in place with his left foot. He uses a splitting knife and mallet to split the log repeatedly in the middle. The final split requires especial care, the splitting knife has to be placed very precisely, it is tapped in only lightly, and then placed diagonally in the cut in the beam. The split opens up with a light twisting and levering movement, the knife goes in more deeply, until finally the two shingles can be drawn apart with both hands.

The shingles are now placed individually on a trestle (*kezuri-dai*), a thick plank set up diagonally with a few laths, on to which a stop has been nailed. The craftsman stands behind the trestle, which comes up to his chest, and pushes his long knife down diagonally to taper the shingle; this takes off a long shaving that rolls up immediately. At the top end (*shiri*) the shingles are only half as thick after this step. If thick shingles are

needed (*tochi-buki*), they are all tapered with the long knife (*shiri wo otosu*, working the rear part off); for thin shingles (*kokera*), which are more frequently used, only the short shingles (*noki-ita*) stacked high by the eaves are tapered.

Covering with wooden shingles

Once the support structure in the form of close-spaced battens or boarding has been put in place, covering starts at the eaves. The studio at the Reiun-in, a subsidiary temple of well-known Myoshin-ji Zen temple in north-western Kyoto was re-roofed in 2003. It is a late 16th century building on an almost square ground plan, with a hip-and-gable roof. The shingles were stacked above each other in many layers at the eaves, a process known as *noki-tsuke*. They consist of a lower layer, *ura-ita* (l. 240 mm, t. 18 mm), eight layers of eaves shingles (*ko-noki-ita*; l. 150 mm) and above this one or two layers of *uwame-ita* (t. 3.6-4.5 mm). The end grain on the high eaves is trimmed with a plane, having first been moistened. Once the eaves have been built up, the roof diagonal can be covered, *hira-buki* (flat covering). The shingles are 30 cm long, 15 cm wide and 3 mm thick, and are laid in rows (*fuki-ashi*) 3 cm apart, the distance marked out with a snap line. Here the string is not drawn through ink, but through water-soluble red paint (*shu-iro*). The shingles are fixed with bamboo nails. These are 3 cm long, 3 mm in diameter and are made of the hard edge of the giant bamboo (Môsô-chiku; Phyllostachys pubescens). The nails are heated in a large pot to draw out the sap contained in the bamboo. This process also makes the nails harder and more durable, and their surface smoother. This is also why the roofer is able to carry 30 to 50 such nails in his mouth between teeth and cheek without getting splinters. He twists the individ-

Building site at the Myoshin-ji.

The shingles are fixed to the roof boarding or battens with bamboo nails. The slight irregularities in the shingles create air channels that enable the roof covering to dry out quickly.

The roofer takes a nail out of his mouth, puts pressure on it with the metal strip fastened to his hammer handle, then drives it in.

ual nails with the tip of his tongue so that the tip points inwards and forces one out of his mouth. He presses on the shingle with his left hand, takes the nail with the thumb and index finger of his right hand, in which he is also holding the hammer, then presses the nail on to the shingle with the small iron plate attached to the hammer handle. Now he drives the nail in with three or four blows. These blows mean that the bamboo nail splays out a little at the head, and is thus thicker, which helps to keep the shingles in place. The bamboo nails last for an astonishingly long time.

Particularly elaborate work is needed at transitional points, as here between a roofed corridor and a hall.

Bark roofs

Bark roofs are found, above all, at Shinto shrines and some Buddhist temples; until the early 20th century they were also very occasionally used for private houses. By far the largest ensemble of bark-roofed buildings is the Imperial Palace in Kyoto, Kyoto Gosho.

Here a roof area of 15,000 m² is covered with bark. The roof covering is renewed every 30 years, usually as part of a long-term campaign. The present campaign runs from 1988 to 2012, and involves craftsmen from eight workshops.

Harvesting the bark

Cypress bark is used for the purpose. The trees grow from the main south island of Kyûshû to the Fukushima Prefecture in the north of Honshû. Their wood is fragrant, and has a high resin content, which also makes it ideal building timber. The bark of Japanese cedars is also used for shingles, particularly for teahouses, but it is less durable, and is used much less. Unlike cedars, whose bark is removed after felling, cypress bark is harvested from the standing tree, in the second sixth months of the year. If it is removed carefully, cambium and bast are not damaged, and the tree grows new bark about 2 mm thick within ten years, and this can be harvested again. This newly formed bark (*kurose-gawa*, bark with a black back) is particularly sought after; it soon goes darker in colour and is smooth, and thus easy to work with. When bark is removed from a cypress for the first time it has deep cracks, contains little oil and has a rough surface (*ara-kawa*).

The ideal bark comes from cypresses over a 100 years old, which are stripped at regular intervals. But the lack of such stocks has meant that bark is harvested from trees that are only 60 years old. Trees on windy ridges are not suitable, neither trees that are over 300 years old, as they increase in thickness too slowly, and are difficult to climb. The bark collectors (*moto-kawa-shi*) have three main tools: a rope 20 m long and 2.5 cm thick (*buri-nawa*), whose ends are tied to 50 cm long 3 cm thick staves (*buri-bô*) made of hard Kashi wood, a 40 to 50 cm long wooden scraper (*ki-bera*) and a kind of machete (*koshi-bôchô* or *nata*) for cutting the bark off and into lengths.

The bark collector climbs the tree with the aid of a rope and two wooden rungs fastened to it, then loosens the bark up to the lowest branch.

The pieces of bark are trimmed in the workshop; some are split and "clipped together."

The bark collector introduces the tip of the wooden scraper vertically into the bark and pushes it inside the bark for about 15 cm. Then he thrusts the scraper strongly against the tree, so that the tip breaks through the bark from the inside to the outside. Now he can hold the scraper with both hands and carefully separate the bark from the tree. If it is loose at the foot of the tree he grasps the strip of bark with his left hand and detaches it with the scraper to above head height. Finally he takes hold of the bark with both hands and pulls it off to a height of 3-4 m. He does not step far back from the trunk when doing this, as otherwise the strips would get thinner and tear. The bark is detached with a wooden scraper that the bark collector carves from a piece of split Japanese photinia (*kanamemochi*; Photinia glabra), which is highly thought of for its hardness and toughness. A bark collector can harvest from five to ten trees per day. He puts the strips one on top the other, ties them together and cuts them to length, usually 75 cm, with a large knife (*ôkiri-bôchô*). The trade standard is 30 kg bundles, called *maru* (round).

Making bark shingles
Pieces of bark of different shapes and sizes are used for roofing. They can be divided into three groups: bark for eaves (*noki-kawa*), for the roof diagonal (*hira-kawa*) and special shapes (*dogû-kawa*, "tool bark") for uses like verges, hips, valleys and curved entrance porches. The bark is usually prepared in the workshop (*koshirae*). The roofer sits cross-legged in front of a round timber about 30 cm high, usually a piece of cypress or pine trunk. He moistens the top of this rest, called *ate*, about every half hour, which protects his tools. The most important tool is the bark knife (*hiwada-bôchô*), which has a blade about 35 cm long with two cutting edges, one long one and a short one on the oblique head. The back of the blade is slightly concave and comes to a point at the head.
The first step is the cleaning of the bark (called *arai-kawa*, washing bark). Faults and damaged areas are cut out, the bark is smoothed a little if necessary and particularly thick pieces are split into two or three layers, each just under 2 mm thick. To do this, the craftsman taps lightly with the back of the knife on the end of the bark, which has been placed on the block, pulls it slightly apart with his hands, introduces the knife into the crack and opens the bark up. The strips for the eaves are only 3 cm wide and come to a point at the top. Hence this smallest format tends to be made from left-over pieces. The bark shingles for the roof diagonal are considerably more elaborate, and make up 90% of the total. They are usually 75 cm long, 10 to 15 cm wide and between 1.5 and 1.8 mm thick. These shingles taper at the top to about a third of their width. The majority is assembled from two or three narrow strips (*tsuzuri-kawa*). The strips are laid so that they scarcely overlap at the lower end and by about 6 mm at the top. The roofer taps lightly on the overlap with the point of the bark knife, so that the upper layer penetrates the lower one like a thorn and holds the two together. Even more care is needed for the numerous special shapes, even though they form only a small proportion. As the verge, hips and valleys are particularly sensitive, and also particularly conspicuous, specially good bark with no cracks is used. The lower edge of the hip shingles is rounded, while the valley shingles are cut slightly concave.
The completed bark shingles are piled up, bundled and taken to the building site. Preparing the bark takes about three times as long as actually fitting it. So in many firms the older workers prepare the bark in the workshop while the young ones do the roofing on distant sites.

Two narrow strips are put together and fastened at the overlap point. This work is done on a trunk section that is regularly moistened.

The lower edge of the bark shingle is cut.

Roofing with bark shingles

Once the battens, or in very curved areas of the roof the boards, are in place, the roofer starts work on the eaves (*noki-zuke*). The building can have single or double eaves, here the bark strips or timber shingles are layered in a complex structure up to a thickness of 60 cm. At a first glance it seems as though the whole of the roof is covered to this thickness, but the actual roof area is usually barely 10 cm deep. The thick roof edges are a result of the two-shell roof frame used since the 10th century for religious and public buildings.

A long transverse timber is nailed a little way back from the ends of the rafters, *kaya-oi* ("supporting the reed", the somewhat confusing name for this component is independ-

Bark formats –
from left to right
noki-tsuke
(for the high eaves),
tome-kawa
(small and particularly thick piece of bark for the corners of the roof),
uwame
(laid over the sheet copper),
hira-kawa
(bark pieces cut slightly conically for the roof diagonal),
hafu-nama
(short, conical pieces of bark for the gables),
sumi-kawa
(triangular pieces for the hips),
tani-kawa
(for valleys).

ent of the actual roofing material). A layer of short thick planks (*urago*) is fastened on top of this. The bottom layer of the eaves is made up of thin, short boards (*ura-ita*) or strips placed diagonally (*jabara*). Strips of bark 3 cm wide and tapering to a point at the back are laid on top of this in up to 100 layers. Every few layers a strip of wood with a wedge-shaped cross-section (*dôbuchi*) is placed about 15 cm further in. These strips stabilize the eaves: it is not possible to layer to a height without them, and they create the air channels needed for back ventilation. Once layers of bark have accumulated to the required thickness, a strip of sheet copper is fixed at the top. It protrudes by about a centimetre at the front and is folded down so that the water can drip off thereby protecting the eaves. Then come a few layers of relatively large pieces of bark cut to shape at right angles (*uwame-gawa*, bark on the top), before the shingles are laid in steps on the roof diagonal. The distance between the rows increases from the eaves to the ridge of an increasingly steep roof from 6 to 15 mm, and so the covering becomes thinner towards the top. The strips of bark are fixed with bamboo nails about 35 mm long with a cross-section of 2 mm.

Several young roofers
working in a row, sitting on triangular
stools nailed together
from battens.

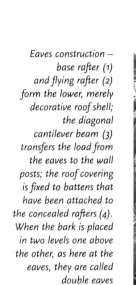

Eaves construction – base rafter (1) and flying rafter (2) form the lower, merely decorative roof shell; the diagonal cantilever beam (3) transfers the load from the eaves to the wall posts; the roof covering is fixed to battens that have been attached to the concealed rafters (4). When the bark is placed in two levels one above the other, as here at the eaves, they are called double eaves (futa-noki).

Many layers of narrow bark strips make up the 20 cm high eaves. The copper sheet strips laid above them with drip mould make the eaves last longer.

Covered building site at the palace. The support structure battens were partially replaced.

At the eaves and on the verge the strips of bark, piled high on top of each other, are sprinkled with water and trimmed with an adze. At the top of the gable in particular, patterns can be worked in by shifting levels. The roofer's skill is demonstrated particularly at the eaves, where precision is most in evidence.

Bark roofs last for a good 30 years, thus a little longer than wooden shingle roofs. They last for longer if they are constantly maintained after storms or damage by crows. However, high costs have meant that the elaborate and more durable eaves ends are not renewed in every campaign.

The future of roofing

There are about 100 roofers in the whole of Japan who work with bark and timber shingles. The number of buildings roofed with these is estimated at 3000, and just under half of these are state-listed as protected monuments. So the profession survives mainly through constant demand generated by monument preservation. Increased earnings and publicly funded courses have helped to reduce the average age of the roofers considerably. But there is concern about getting hold of suitable bark. It was originally a common sideline for farmers and forestry workers in the six winter months, especially in the Tanba region in the Hyôgo Prefecture. In the first half of the 20th century there were about as many collectors as roofers, but only about 20 bark collectors are working at the time of writing. Low incomes and increasing distances to travel have brought down the numbers in this profession, which is dependent on the weather and not without its dangers. Roofers have sometimes been forced to use bark from felled trees (*doba-muki*, stripping on site), but the durability is a third less. Acquisition is also made more difficult by a lack of suitable trees. The cypresses should be at least 80 years old, but as forestry cycles have shortened, there is little old tree stock. Bark collecting and roofing used to be separate trades, but for a few years now some roofers have been going over to collecting their own raw material. This includes Sugimoto, who has had three of his employees collecting bark in private woodland in the prefecture since 1998.

Bark shingles on a hip in the roof.

96

The bark roofs of the Imperial Palace in Kyoto are often topped with tiles at the ridge. The bark shingles, piled high at the eaves and verge, give the impression that the building is protected by a thick cushion.

内装と家具

Interior and furniture

House altars
Shôji – sliding doors
Ishô-dansu – trousseau cupboards

Fitting out a house altar with liturgical equipment, here following the example of the Hongan-ji branch of the New School of the Pure Land. A lectern for Sutras, a singing bowl and a censer are among the items in the altar, and there are several small hanging lanterns.

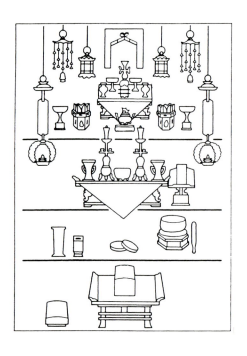

House altars

The Buddhist house altars known as *butsu-dan* are to be found in almost every Japanese home, mostly in a niche or wall cupboard. From the outside they look like large cupboards themselves, but their elaborate interiors are reminiscent of temple halls and appear very costly. Inside they are fitted with all sorts of equipment, a small lectern for the Sutra book, a singing bowl, hanging lamps. Usually there are small, narrow tablets on the shelves with the names of the dead engraved on them. These indicate the most important function of the house altars: preserving the memory of the family's own ancestors. A small offering of food is made here several times a day, before breakfast and the evening meal, usually a little rice. There are also occasions on which prayers are said annually before the altar, at New Year, for example, or the August days in celebration of the dead. On the anniversary of a family member's death a priest is often called to read Sutras at the house altar in exchange for a donation.

Little altars for personal use by the nobility and priests existed even in the Middle Ages, but house altars in their present form became generally available in the 17th century. Under the Tokugawa-Shogune regime, every household had to register with a temple. Setting up a house altar showed the family's affiliation to a particular school of Buddhism. Given this, building Buddhist house altars came to be a profession in its own right.

Typical house altar from Kyoto – it is lacquered on the outside, the highly decorated inside is gilded.

Numerous variants have developed in the last 400 years, typical of a region or a sect. Roughly speaking, a distinction can be made between three basic types today. The traditional version is painted black and gilded on the inside (*nuri-butsudan* or *kin-butsudan*), with the body and interior made of coniferous timber most of the time. In the late 19th century house altars started to be made of expensive imported timber like ebony or rosewood (*karaki-butsudan*), in some ports, chiefly Tokyo and Osaka. A third type has emerged in the past 20 years. It is considerably smaller, and more reminiscent of designer furniture from the outside (*shin-butsudan*). This is essentially because families are getting smaller. They do not want to give up the usual location for domestic ancestor worship, but there is not much room for an altar in their cramped homes. These modern house altars are thus particularly in demand in big cities.

Complete but uncoloured altar for a temple. It has lavish base and cornice mouldings, and a carved balustrade fitted at the top.

Regardless of the nature and quality of the house altar, the interior is always lavishly decorated with carved ornaments, lacquer paintings or a large number of decorative fittings, which are gilded like the inside walls. As a space for the Buddha the interior is not of this world, but despite its function, directed at the next world, the house altar is not infrequently also a status symbol for the owner, who likes to use it as a sign of affluence. This is particularly noticeable in the country, where farmers who have become rich by selling land like to set up a pompous house altar.

Building house altars is definitely a distinct branch of commerce in Japan.[1] It employs about 12,000 people in the country as a whole. One of the major altar-building centres is the old capital, Kyoto, with over 1000 craftsmen. House altars made here are called *kyô-butsudan* after the city, and are all painted black on the outside. One special feature is that the Kyoto altars are made with divided labour and still by hand in over 300 workshops, most of them small and highly specialized. The house altar businesses are mainly near large temples. This includes the altar builder Kobori, whose main shop is just opposite the huge Higashi-Hongan-ji temple. When family members die, the most frequent reason for buying a new altar, the remaining family often seek out the main temple of their confession; afterwards they frequently enquire about a suitable shop. So being close to a temple and enjoying its trust is crucially important to the altar-builder. Kobori sells about 800 altars per year in his five branches, with prices ranging from € 5000 to over € 300,000. Kobori works like a publisher, and has parts made in 150

Shelf with a large selection of moulding planes, as still used for the lavish altar mouldings.

House altars

The balustrade from the stockroom is fitted into the central section of a house altar; it is scarcely 6 cm high, Ueda's father takes almost two days to carve a balustrade with lotus motifs like this one.

small workshops. There are workshops for the elaborate little roof that adorns the interior of the altars, for carved parts, for the door latticework, for decorative fittings, framing, mother-of-pearl inlays and gilding. All the components are then assembled in his own workshop, which employs painters and gilders as well as cabinet makers. The only items the workshop is solely responsible for are altars and shrines for temple halls, which are the business's second key stand.

Kobori likes to have the wooden core (*kiji* – the same term is also used for the core of lacquered vessels) for house altars of particularly high quality made in the Ueda workshop; he commissions three to four wooden housings from this two-man business every year. Ueda uses only very close-grained timber because the cross-sections are so thin, for door bars and the interior of the altars, for example. Native cypress wood is chosen for the highest grades of work, otherwise Russian or North American pine is used nowadays. From the outside the body of the altar resembles a piece of furniture, with its four corner posts and panels set in grooves, its base and cornice mouldings and revolving doors at the front. The central inner section looks like a Buddhist temple, however. This link with architecture for places of worship is particularly clear when the altar is being made for a family belonging to the New School of the Pure Land (Jôdo-Shinshû). In this case the interior is modelled directly on the central hall of the main temple. House altars are made in standard widths from 60-120 cm. Ueda makes a lot of components to keep in stock, like carved panels, the little balustrades or the intricate coffered ceiling. The storeroom on the upper floor of his house makes him better able to cope with urgent orders. Plywood has been used for the panels for about forty years; it is easy to colour and gild, and it also avoids the danger of dry cracking which occurs as the altars are often kept in air-conditioned rooms nowadays. Apart from this and the use of woodworker's white glue rather than rice glue, the work is carried out on traditional lines.

The Japanese house altar market has had a shock. House altars have recently started to be imported from China, commissioned from there by a Japanese wholesaler. Kobori does not intend to stock them, even in the future, though he does have one of these

Ueda has a storeroom on the top floor of his house containing finished components he has made himself for house altars.

Ueda father and son with the central part of a house altar. The columns and the five-stepped bracket complexes make it clear that elements of sacred architecture have been taken over here.

altars in his workshop – so that he can point out the differences in quality to his clients. Persuading customers about quality seems the only possible strategy in his situation: the Japanese altar builder cannot compete with the Chinese imports in terms of price, as his products cost about seven times as much. Kobori, who makes particularly elaborate altars and house altars, is not yet suffering from the imports, but other manufacturers of simpler models are already finding that their sales are collapsing.

Detail of the columns with decorative fittings. They are still beaten and chased by hand for high quality pieces.

Shôji – sliding doors

Japanese timber buildings are skeleton constructions. The bays in the load-bearing structure can be filled with clay or boards, but moving elements in the form of sliding doors usually closed most of the apertures. These sliding doors are the key to opening up the building to the outside world and to modifying the plan easily, two basic features of Japanese architecture.[2]

The oldest surviving timber buildings like the main hall of the Hôryû-ji and Tôshôdai-ji temples near Nara have relatively few high, two-leaf doors made of thick planks mounted with cleats at the top and bottom to prevent warping. A movement towards ever lighter constructions and greater diversity can be observed from the 9th to the 16th century. *Shitomi-do* are the first development: a panelled frame with a fine-meshed lattice above it is mounted between two posts. *Shitomi-do* are usually in two halves; the lower section, about 80 cm high, was usually fixed with latches and removed only on special occasions, but the upper section was suspended from the lintel rail and could be folded upwards.

From the 12th century onwards, unpanelled lattices were also made, still fixed in the early stages. The oldest surviving example of a moving lattice door that can be slid open is in the Seirei-in hall at the Hôryû-ji temple, which was built in 1284. Here it separates the hall's inner and outer sanctuaries. Originally the groove needed to guide the sliding door was created by nailing thin battens to the lintel and the sill. *Shôji*, covered with transparent paper on one side, and *fusuma*, which are covered with several layers of paper on both sides, are the two extremely light sliding door variants that are now associated with traditional Japanese architecture; these emerged at the turn of the 13th to the 14th century. The first *shôji* were installed directly behind the board doors of the

Inside rooms in the Reizei villa, built in 1800, immediately north of the Imperial Palace in Kyoto. The sliding doors can be pushed back or easily removed for special occasions.

Interior and furniture

104

outer walls, with several doors running in a single groove. Two grooves became customary at a later stage, and *shôji* moved to become room dividers inside the building. Numerous new variants emerged, and *shôji* with a solid panel in the lower section started to occur frequently. These were called *koshi-tsuke shôji*, a *shôji* with a hip.

The first pictures of sliding door makers date from the late 16th century, and by the end of the 17th century there are records of a door-building quarter in Osaka, indicating that this was a frequently occurring profession. But extremely high-quality work on 14th and 15th century doors suggest that there were craftsmen specializing in their manufacture at a much earlier date. From the early 17th century the moving parts of the buildings are called *tate-gu*, a composite of the verb *tateru* for construction and *gu* for equipment/ furnishing. So the profession is called *tate-gu-ya* or *tate-gu-shi*.

With the exception of two niches, monument preservation and teahouse building, sliding doors in Japan are now machine made, and sometimes even completely automatically.

Ink marking starts with laying out and marking the planed battens.

The Suzuki workshop

One of the very few joiners who still build a wide range of hand-made doors is Suzuki Tadashi in Kyoto, who was born in 1936. Suzuki was apprenticed to a well-known joiner who finally adopted him, a frequent form of succession in Japan. Over the past 50 years he has been involved in restoring some of Kyoto's best-known buildings, including the Katsura Villa and the Phoenix Hall of the Byôdô-in on the Uji river. His mastery and also the importance of his profession were publicly acknowledged in 1999, when the government recognized him as a representative of a craft essential to the maintenance of ancient monuments. Since then he has drawn a small annual scholarship, and with the Kyoto Prefecture offers an annual seminar for joiners working in monument preservation.

In order to be able to record the most important manufacturing stages, a sliding door was commissioned from Suzuki. Its dimensions and construction are identical with the door for a tearoom at the Gyokurin-in, one of the many subsidiary temples of the well-known Daitoku-ji Zen temple in north Kyoto. This tearoom is called Kasumi-doko. It has an area of four and a half mats, and is known for its unusual alcove, whose two staggered shelf floors are suspended freely. Suzuki had renovated two sliding doors during the 1977-78 restoration, and so the door that was ordered was extremely familiar to

Marking is done with a single – or in the case of the recesses in the bars – a double knife. Here the double knife is being adjusted to the thickness of the bars.

A set-square made of a thin hardwood board is used as a stop.

The tenons for the bars are marked with a small, home-made gauge.

him. Making a sliding door by hand as described here would be an unusual commission for most joiners, but it is an everyday matter for Suzuki. He is able to demonstrate techniques that are hardly used any more as part of a unique routine.

Suzuki cuts the recesses in the bars with a dovetail saw to which he has attached a depth stop.

Choosing the timber and marking out

Cypress is the preferred wood for frames and bars. It comes from trees that are several 100 years old; it is dead straight and astonishingly light. Suzuki buys it in rails from a Nagoya timber merchant who specializes in Hinoki wood from the Kiso region that is especially sought after because it is so close-grained.

The two upright members, the stiles, are called *tate-gamachi*, and the three cross-members are called *shimo-zan* (bottom rail), *naka-zan* (middle rail) and *kami-zan* (top rail) according to position. For this sliding door, a Japanese cedar panel is set in grooves in the bottom third, with another bar (*yoko-maira-zan*) tenoned in halfway up the panel. The lattice in the top section is called *kumiko*; here it consists of three upright and eight horizontal thin bars (*tate-kumiko* and *yoko-kumiko*). The wood is assembled so that the straight mild grain is at the front and the back. Attention is also paid to the fact that surfaces that are frequently touched in use and for cleaning purposes do not display the heartside to avoid splinters. So this side is on the outside for the two uprights and on the top for the bottom and middle rails.

Cutting (*kidori*) and planing (*ara-kezuri*) are now done by machine, even for sliding doors for ancient monuments. But the subsequent marking (*sumi-tsuke*) and further

The small tenons for the bars are cut with a dovetail saw and trimmed with a rebate plane.

working processes are all done by hand. First the stiles are marked together, so that they can be easily identified during later work and the assembly phase. The markings are done with a red marker; the joiners call them *ai-jirushi*, signs of belonging together. The stiles are marked with a gauge (*wari-tsuke-jôgi*), a thin wooden bar with all the important measurements on it. This gauge is particularly useful when several sliding doors are being made, or if follow-up orders are placed, as the dimensions can be easily transferred on to the battens, which are held together with little clamps. Knives are always used for marking across the grain; a blade sharpened on one side (*hira-gaki*) for single marks, a double marker knife (*nichô-hira-gaki*) for two marks. In the latter case the two thin blades are held together at the top by a rivet, and two screws lower down allow the distance between the blades to be adjusted. The blades are always sharpened on one side, with the cutting edge on the outside. A set-square (*maki-gane*) made by the joiner himself from a small hardwood board is used as a stop. This square can be easily trued up with a plane if necessary.

The thin bars for the lattice are marked out after the stiles. Suzuki drives some thin nails in at the top to hold the bars together. He marks the recesses with the double-bladed knife.

Wood joints

The recesses in the bars are cut with a dovetail saw (*dozuki*), to which Suzuki has attached a depth gauge with two iron clamps. He makes three incisions at each recess before removing the wood with a tool he devised himself. It is made of a piece of an old saw blade, with a wooden batten attached at the top. The blade is introduced into the saw cut and pushed sideways. The wood then separates at the bottom of the saw cut and can be removed easily. The bottom of the recess is cleaned with a special cutting tool. Then the little tenons on the bars are cut (*kuminoko no hozo-tsuke*); the shoulders of the tenon are offset and the battens cut down to their final dimensions

before the actual tenons on the battens, which are held together with small clamps, are cut with a rebate plane, first taking off thick shavings with the large rebate plane (*dai-ara-shakkuri*) and then switching to the small rebate plane (*sho-shakkuri*). Suzuki now removes the clamps and turns every second batten over and checks whether the wood is slightly discoloured at any point. If it is he will arrange it so that that the discoloured parts are on the side that will later be covered with paper, and so not attract attention. Incidentally, the small tenons on the bars are slightly chamfered with the plane on all four sides (*hozo-men wo toru*), an indication of the great care and precision that go into the work here.

Suzuki now goes back to the frame members, and cuts the tenons on the rails. He made the little marking gauge (*ana-kebiki*) himself, from Kashi wood. As for most of his tools, these marking gauges have their own wooden box with a sliding lid, containing large numbers of them. The many cutting tools are also treated with particular care, with their delicate blades each in its own specially made wooden sheath (*yôjo saya*).

Next the mortises for the lattice are cut out (*kumiko no anabori*). The wood shavings from the narrow tenon holes are taken out with a *mori-nomi* or harpoon cutting tool. The bottom of the mortises is cleaned with the *soko-sarae*; this special L-shaped cutting tool is reminiscent of a router plane in shape and effect. Suzuki uses a long wild

Frame members with
completed corner
joints; the edges of
the tenons are
slightly broken.

Isometric drawing of
a corner joint.

cherry plank as an underlay when mortising (*kezuri-dai*); it is also used for planing and jointing.

Suzuki now turns to the four corner joints for the frame. Here the top and bottom rails are jointed on with a small gooseneck tenon. This elaborate joint was still being used on doors in the early 17th century, but then only for very high-calibre buildings. The oblique surfaces of the trapezoid gooseneck tenon are cut with a knife and a little plane. The mortises need special care, as they have to correspond precisely with the oblique shoulders of the gooseneck tenon. Suzuki has made a little template for this; he puts it in the mortise and can move the cutting tool along it.

Chamfering bars and making the panel

Work now continues on the lattice, whose bars are chamfered. This step is made more difficult by the fact that the bars are "woven", so the recess is alternately at the front or the back. Suzuki cannot just run the plane along the bar, he always has to stop before the cross halving. First of all he lightly bevels the edges with a small compass plane (*sori-dai*), then the takes the chamfering plane (*sarubo-mentori-ganna*) and smoothes the chamfer. Finally the panel for the bottom section of the sliding door is made, in this case by gluing five thin boards together. As for the original, a V-shaped joint is made; two special moulding planes (*ken-ganna*) with two side fences are used for this. In former times, butt joints were avoided for panels assembled from several boards or for large boarded areas; rebated or tongue-and-groove joints were preferred.

Assembly

Before the door lattice is woven and glued, Suzuki taps the wood at all the crossover points with a small hammer, and in each case on both sides of the bar, at points that will be covered in assembly. This is a technique used in many timber-related professions for making joints easier to assemble, and is called *ki-goroshi*; the literal translation, "beating the wood to death", sounds violent, but the hammer blows are careful, and meted out lightly. The compressed wood will swell back into shape after the glue is applied, and the joint will be absolutely secure.

Shōji – sliding doors

The bars are put in a block and the edges chamfered.

The wood is tapped lightly with a hammer at the recesses; this makes it easier to fasten the bars together.

Weaving the lattice – the recesses are alternately on the top and the bottom.

The structure used for the lattice here is not called *jikoku-gumi* or hell's weaving for nothing. Assembly needs a special technique, and maximum concentration. Suzuki applies glue to two of the three upright bars and the two corresponding horizontal bars. Then he puts the two upright bars on his bench and adds the first to eighth horizontal bars one after the other. He then applies glue to the rest of the crossover points. He sets the half-assembled lattice upright and carefully prises the transverse bars apart with two long wooden battens, a thicker one near the tenons and a thinner one underneath the recess. Finally the third upright bar can be worked in, and the two pieces of wood used to prise the structure apart can be taken out. This is a risky step, as the bars can break or split if too much pressure is applied. The concentration needed is correspondingly great, as an uncontrolled movement can ruin the whole job. When the lattice is woven, the joints are checked for flushness and knocked into place with a hardwood block if necessary.

When the lattice glue has set the cross halvings are smoothed with a small plane, thus removing minimal differences in height caused by the timber swelling (*me-chigai-kaki*). Then the chamfers on the halvings can be dealt with. Here the chamfer plane had been stopped so that the chamfer does not run through the halving. Suzuki places the lattice

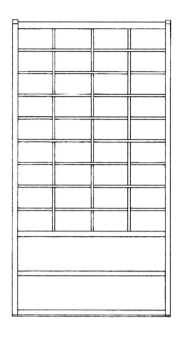

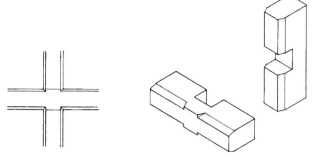

*Drawing of the sliding
door with details of a
bar intersection.*

*The chamfering is cut
with a knife at the
lattice cross halvings.*

upright on his bench and cuts the chamfers around the cross halvings to shape. For this he uses a special pointed knife with cutting edges on both sides (*kiri-tome-bôchô*).

A rebate of only 0.2 mm for the paper covering is now planed into the frame members (*kami jakuri*). As in the case of the chamfers on the bars, this rebate cannot be continuous, but has to be interrupted. Finally the frame members can be hand smoothed and the internal edge lightly bevelled.

The panel is prepared to the appropriate width. The joiner does not use a saw for this, but a large cutting gauge with a protruding blade. He runs along the edge of both sides two or three times with it and then works the thin board. This tool is called *wari-kebiki*, literally a wood marking gauge for cutting. The thin board is cut to shape with a dovetail saw heightwise, in other words across the grain. The panel is then hand-finished, first with an intermediate plane (*chû-ganna*) and then with a smoothing plane (*shiage-ganna*). As a preparation for building it in, the edges are not just worked, but here again

*The Japanese cedar
panel is cut to size
and smoothed. The
iron blade has to be
very well sharpened
for the soft wood.*

Shôji – sliding doors

the wood is slightly compressed. This is not done with a hammer, as for the bars. Instead the above-mentioned cutting gauge is adjusted so that the distance between the tip of the blade and the stop is exactly the same as the thickness of the panel. Suzuki then runs the cutting gauge on edge round the panel from both sides, thus slightly compressing the edge of the thin and relatively soft Japanese cedar board.

Before the trial assembly of the frame members, Suzuki also compresses the rail tenons with a few hammer blows. Then he takes off his slippers, climbs on to the bench, holds the stiles with his feet and carefully knocks the top and bottom rails in. The actual assembly takes place in stages. First the lattice is tenoned into the middle rails. Then Suzuki places the panel into the groove of the centre rail and tenons the two stiles on,

Test assembly of the frame – Suzuki takes off his slippers and climbs on to his bench.

Assembled sliding door – one split bamboo bar is still being attached to the panel.

Finishing – minimal differences in height are removed with a small plane.

before the top and bottom rails can be pushed into place. These corner joints with the gooseneck tenons are so precisely cut that he does not need to apply any glue; this makes it easy to dismantle the sliding door for any repairs that may be necessary later. The parapets are carefully moistened with a toothbrush and the rebate for the paper covering is cut into the corners. Finally the stiles have a shoulder cut into their ends so that they fit into the groove of sill and lintel. The stiles will be allowed to protrude a little above the rail at the top end.

Interior and furniture

112

In the sliding door described here the stiles and rails are the same width when installed, and so the door looks like a light picture-frame. In historically later versions the rails are often considerably wider than the stiles, so the doors look heavier.

The strict form and precision of the sliding doors is reflected in the way Suzuki works: he concentrates very hard, and works quickly without seeming to rush. His little work-shop is also appealingly tidy: when he has finished each stage of the work he clears away the tools he no longer needs and cleans his workplace. So the five-day visit to his workshop was both calming and uplifting for the viewer.

Final step – the two stiles are recessed at the top so that they fit into the lintel groove.

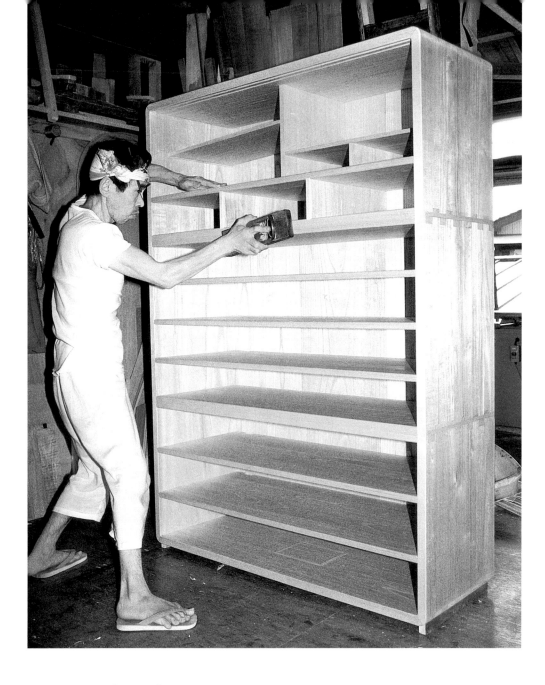

Ishô-dansu – trousseau cupboards

There are numerous variants on traditional chests of drawers for clothes. They are mainly drawers of various heights and widths that slide in and end up flush with the front of
the furniture, though sometimes they can be hidden behind sliding or revolving doors.
The divisions become more intricate towards the top, the drawers decrease in height
and two, three or four small drawers can take the place of a single wide one. The chest
of drawers used for clothing usually consists of three elements on top of each other
(*kasane-tansu*), similar to a type of storage furniture, the so-called *Brandschränke* found
in Germany and consisting of stacked chests with handles that could be quickly carried
out in case of fire. In Japan, there is always a hook at the top of the sides that can be
pulled up to take a wooden pole for transport purposes. The numeral for *tansu*, *sao*, is
also linked with this idea of transport; on its own it means pole.

The first small piece of furniture with four drawers was found among the 8th century
objects kept in the Shôsôin, but this idea seems soon to have been forgotten again.[3]
The next pieces of furniture with drawers did not appear until the late 17th century.
They became common in the 18th century, and were made in many variants, to hold
swords, for example, documents, medicine or as a kind of safe for shipowners. Some
had wooden wheels so that they could be moved easily in case of fire.[4]

114

These chests of drawers were known as *tansu* from the 18th century, and their external parts were originally made of Keyaki, a hardwood similar to our elm. Kiri was used only for the insides. Then in the early 19th century cupboards started to appear made entirely or largely of Kiri. They were very popular and soon became a fixed dowry item for daughters from better-off households. People were starting to own more clothes because of increasing affluence, which meant that this piece of furniture caught on. Several manufacturing centres came into being, the best-known being the little town of Kamo in the Niigata Prefecture, not far from the Sea of Japan. This was a good place for cabinet-making because there were large stocks of good Kiri wood at first, and the neighbouring town had a flourishing tool-making industry.

Construction

Tansu are cubic pieces of furniture. Board construction is used particularly for trousseau cupboards, which are mostly made of Kiri. Sides, bottom and top of the carcase are usually joined with finger-laps or through dovetails. These corner joints are additionally secured by wooden nails. The intermediate shelves between the drawers are also made up of boards; these are set in grooves or dovetailed and are fixed at the front with wooden nails driven through the sides of the carcass. To save material, the intermediate

On the inner corner of the carcass a curve is cut corresponding with the outer one. A thin batten is mitred for this purpose.

Placing the short battens in the carcass corner.

The carcass is thoroughly rounded off at the top, at the point where the sides meet the lid.

The drawer floors are glued on; here the son is pulling the thin floor on to the rebated front piece with a clamp.

Box with conical wooden nails.

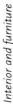
Oyanagi Senior has placed the carcass on his working surface so that he can plane it smooth.

shelves reach their full thickness only at the front and back, they are thinner in the middle. The drawers are simply constructed. The sides are only rarely connected to the drawer front with lapped dovetails, usually the drawer front only has a simple lap and the sides are fastened with wooden nails. The sides and back are joined by some wide finger-laps, and here too conical wooden nails secure the joint. A thin bottom is glued and nailed to the drawer back, sides and the front, which is rebated for this purpose. The grain of the drawer bottom is at right angles to that of the intermediate shelf, so when it is being opened and closed the drawer bottom slides directly over the intermediate carcass shelf. This primitive construction without any ledges for guidance only works because the drawers are light, and fit precisely into the aperture intended to receive them. This precision is clearly in evidence when a drawer is closed quickly; the adjacent drawers are forced out because the air cannot escape fast enough. The cupboards often contain little secret compartments, the preferred positions for these being the plinth of the bottom element and the space between or behind the smaller drawers. Almost all *tansu* have the thinnest possible cross-section, resulting in extremely light and mobile containers.

It is not easy to take thin shavings off the soft and often tricky grain Kiri wood.

Material

The trunks are felled shortly before the start of winter or immediately after the snow melts, and are 20 to 50 cm in diameter. They are cut in a local sawmill in which all the local cabinet-makers have had shares since it was founded in 1908.[5] The thick trunks in particular are cut into several segments and then into thin planks with a band-saw to obtain as many straight grain boards as possible. As is the case with other crafts that work with Kiri, the timber is weathered in the open air for about four months after being cut.

Boards with vertical annual rings are in particular in demand for the fronts; they last well and underline the orthogonal structure of the piece of furniture. In the early 20th century, when dead straight markings on the boards increasingly became a synonymous with high quality *tansu*, and the demand could not be met with local timber, the practice of "correcting" the grain began. The edges of the thin boards were worked with a curved hand plane until they followed the annual ring precisely. Then a kind of wood marking gauge with the drawing needle replaced by a blade was used to cut and separate the upper and lower sides of the material. The narrow strips produced in this way were finally glued together to make a plank with "straight mild grain". Imported wood has been used increasingly since 1965, mainly from South-East Asia or North America. The price is less than half that of native timber, but Kiri from China and South America is especially soft, with tricky grain and thus difficult to plane, and it is also prone to darkening very quickly.

116

Surfaces

Surface treatment is also important for *tansu* made of Kiri. In the 19th century the furniture was often finished with transparent Urushi varnish, but the following process has gained acceptance since about 1930. First the piece is carefully washed with hot water to get rid of any dirt and so that any dents in the soft wood can swell up again. Then the surfaces are hand-planed before being rubbed down with a brush made of willow roots in the direction of the grain (*medate*). It is then painted with an abrasive (*tonoko*) dissolved in a dark brown solution (*yashibushi*) made of cooked fruit from Japanese alder (Hannoki; Alnus japonica). It is brushed again when dry and wiped with a soft cloth (*tonoko-otoshi*). A second coat of colouring is applied and it is brushed again before the furniture is finally waxed. The two coats of colouring produce an even shade of light yellow and accentuate the grain. The repeated brushing also structures the surface slightly, and finally the wax provides a matt, silky lustre, and protects against dirt.

Actually the advantages of this easily transportable furniture should be particularly convincing in these days of great mobility, but sales have been declining slowly but surely since the war. In the mid fifties there were still almost 200 workshops in Kamo, now only 38 remain, and the trend is still downwards. Traditional Japanese furniture is not in great demand; since 1960 it has been forced out by Western models, and often by furniture made in South-East Asia.

Contrary to European practice, the larger chests of drawers were rarely placed visibly in living rooms. They were usually in wall cupboards or in fireproof warehouses with thick, clay-rendered walls. Furniture made of Kiri takes on a silvery and ultimately dark grey shade over the years. It is sometimes smoothed with a plane and then undergoes the surface treatment described above again, to make it look as good as new.

A chest of drawers consisting of three stacked carcases. The drawers become smaller towards the top.

容器 と 器具

Receptacles and tools

Boxes
Coopers' goods
Turned wooden cores for lacquered bowls
Chip boxes
Sieves and steamers
Kaba-zaiku – vessels made of cherry bark
Spoons
Moulds
Combs
Geta – wooden sandals
Making tools

Boxes

It clearly makes sense to pack valuable and particularly fragile goods in boxes for storage and transport. In Japan these are made of wood, and they are available in a quantity, diversity and quality that is unique in the world. All valuable ceramics, lacquered goods, picture scrolls and fabrics are packed in little wooden boxes.[1]

There are three particular reasons why boxes are so unusually important. Firstly, valuable objects are more at risk than elsewhere because of the hot and humid climate, and natural disasters. Additionally, many things are brought out only on special occasions, even picture scrolls are not on permanent display in the sparsely furnished traditional houses. Otherwise packed up valuables were stored in storehouses fireproofed with thick clay walls (*kura*, *dozô*). Well-to-do families often have hundreds of wooden boxes of various designs and sizes stacked on shelves along the storehouse walls. The tea cult, which started to catch on from the 16th century, made a major contribution to the spread of boxes by triggering a passion for collecting fine tea utensils. Until then only cult vessels at temples and shrines and valuable court items had been stored in elaborate, custom-made boxes. The tea cult also introduced another role for boxes, which went beyond the function of a container: artists signed their work on the lid of the appropriate box as well, collectors put their seal on them and added short notes.[2] In this way the boxes frequently convey the history of an object. Often boxes like these, with their own value as documents, had a box of their own made for them, a box within a box, like Russian Matrioshka dolls. Wooden boxes are a sign of high quality, so people often like to use them for packing as well. It is not unusual for businesses here to enhance the value of a product that does not in fact need a wooden box. Thus today expensive kitchen knives or fabrics for tailor-made suits are often offered in wooden

Matoba sawmill – logs are stored in front of the building, the cut boards are weathered on the roof.

*Quartered trunks wait
to be cut down further.*

boxes, and even homeopathic remedies and department stores' gift vouchers above a certain value are presented in small wooden boxes.

There are still sixteen box-makers (*hako-shi* or *hako-ya*) in the former capital, Kyoto. Before describing how the boxes are made in more detail, here is an introduction to one of the Kyoto sawmills specializing in Kiri wood, and mainly supplying box-makers.

Matoba – a sawmill in the heart of Kyoto

The little Matoba sawmill is barely five minutes from the famous Higashi-Hongan-ji temple, not far from the main station. It was founded in 1910, and until the first band-saw was bought in 1927, ten so-called *kobiki* worked here: craftsmen who cut the trunks to size by hand. The heart of the business is the large band-saw (*obi-noko seizaiki*): the log is fixed to a carriage (*daisha*) that moves forwards and backwards on rails past the band-saw. Matoba started to cut Kiri trunks imported from North America in 1980, before then only northern Japanese timber was processed. The imported logs, acquired through an Osaka wholesaler, cost only about half as much and provide somewhat harder wood. The proportion of imported goods is now up to 90 %. The trunks are usually thin, about 15-30 cm in diameter, and are stored vertically outside the premises for up to six months. There are two possible cutting techniques, yielding boards with horizontal annual rings (*itame-biki*) and with vertical annual rings (*masame-biki*). In the first case the whole trunk is cut with parallel incisions. This means all the planks show a lively pattern of grain, with the exception of one or two in the middle that show straight mild grain.

*The boards are
stacked vertically in
the racks.*

121

A trunk is lifted on to
the trolley and pushed
into the workshop.

With thicker and closer-grained trunks an effort is made to cut as many boards as possible with vertical annual rings, the kind that box-makers particularly like. The trunks are
first cut into four or eight segments, so the log has to be turned several times. For *edo-
masa* ("Edo-straight"; a technique that originated in Edo, later Tokyo) the boards are cut
on a machine that has been retooled as a band-saw by folding up a feed-table; work is
parallel with the edges of the quartered log, which is turned each time after one or two
planks have been cut. For *hon-masa* ("really straight"), eighths of trunks are cut into thin
boards by the band-saw. The advantage is that practically all the boards have vertical
annual rings, but they do have an oblique inside edge; this has to be trimmed, thus
reducing the yield.

After cutting the boards are hosed down so that no sawdust is left on the surfaces. Then
they are hoisted up to the roof of the building by a block and tackle and there stored
on shelves, with all the boards from one trunk together. They remain upright for three
to six months in open-air racks. Weathering (*ama-zarashi*) draws substances out of the
wood; this helps it to last better subsequently and it does not discolour so much.

The Matoba workshop will only sell its timber by the trunk. It supplies about 50 workshops in Kyoto, most of them box-makers and tea utensil manufacturers. The sawmill's
three employees, of whom one sets and sharpens (*medate-shi*) the blades of the large
band-saw, are all about 70 years old. Kiri wood has its own unit of measurement, which
is still in use today, the *sai*. One cubic metre is the equivalent of 570 *sai*. Otherwise the
traditional unit for cut timber is the *koku*, 3.6 *koku* are one cubic metre. The price for
cut Kiri wood is 300 to 1500 yen per *sai*. Boards with vertical annual rings fetch double the price; not so much because it takes longer to cut, but because only higher-quality trunks can be used for this purpose. Anyway cutting accounts for only 10 to 15 % of
the cut timber price. The most frequently ordered thickness is 3 or 4 *bu* (9 or 12 mm).

The trunk is fastened
to the carriage and cut
with the band-saw.

Possible cuts –
itame – edo-masa –
hon-masa.
The aim is to produce
the highest possible
proportion of planks
with vertical annual
rings, especially from
high-quality trunks.

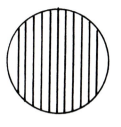

itame

edo-masa

hon-masa

The planks are stored
on edge in racks with
intermediate slats.
Planks from a particu-
lar trunk are always
kept together.

The Matoba brothers,
flanked by two of their
employees.

Boxes

The box-maker's workshop – tools and templates hang on the walls.

Box-makers

Wooden packing boxes are made in a whole variety of qualities. A good box is first recognized by the material; the sides and lid are made from a single board (*ichimai-ita*) with attractive grain, in other words not glued together from several pieces. There are also major differences in the workmanship. High-quality boxes have a gently bevelled or curved lid, and the edges are slightly rounded or bevelled as well. Finally the string (*himo*) also shows whether it is a good piece. If the contents are valuable it is braided by hand even today.

The way the boxes are constructed, leaving no play for expansion or contraction for lid and floor, is greeted with scepticism by Western joiners cabinetmakers, especially for large boxes. But in Japan the timber is so well seasoned that lids or floor bottom split only in rare cases. Often the boxes stand on a little floor frame with small slits in the middle. These are to accommodate the strings with which the boxes are stylishly tied up.

The workshop was founded in 1894, and is now run by Kôsaka Ichirô in the third generation. It is on the eastern side of the old capital city's chequerboard of streets. There were originally a dozen box-makers in the neighbourhood, but only the Kôsaka workshop has survived, specializing in boxes for valuables, and a business somewhere between a home and a workshop next door that makes high-quality disposable packaging for meals (*ori-bako* or *bentô-bako*).

A distinction is made according to use between three particularly common kinds of box: boxes for ceramics, which are usually square (*chawan-bako*), long boxes for picture scrolls (*jiku-bako*) and flat boxes for fabrics (*gofuku-bako*).

Kiri wood (paulownia) is used particularly for boxes intended to contain ceramics and lacquered goods. It is extremely light and durable. As it scarcely shrinks, the containers shut very tightly, which also helps to preserve the contents. If a box is precisely crafted a lid that is carefully put in place usually sinks on to the body of the box in slow motion. Timber pests also avoid well-seasoned Kiri, and allegedly only its surface chars in case of fire. Conifers like cedar, cypress, pine or fir are used as well as paulownia.

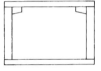
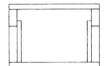

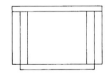

Different constructions for packing boxes.

After gluing, the corner joints and the bottom and lid are secured with wood nails; the holes for these are drilled in advance.

The proud pieces of the wooden nails are cut off with the dove-tail saw.

The rebate is worked so that the lid will fit precisely.

The plane has to be particularly well sharpened to take off only a few hundredths of a millimetre in shavings from this soft wood.

The box is rubbed down with a little fabric sack filled with Ibota-no-hana.

The box is planed smooth on all sides.

Some packing boxes fastened with string – the lids are inscribed in Indian ink and carry the seals of the artists whose ceramics and lacquered goods are in the boxes.

After the wood has been carefully selected, the boards are first planed down and cut to shape, which is now usually done by machine on the ground floor of the workshop. The sides of the boxes are butt-glued or joined with finger-laps. Formerly a glue made from crushed rice-grains was used (*sokkui*), now it is typically white glue. Kôsaka still uses rice glue prepared with a bamboo spatula, but about a third of white glue is added. Glue is applied to the sides and bottom of the boxes; a whole series of them are often piled one on top of the other and fixed into a gluing stand. All the joints are additionally fixed with little wooden nails made of deutzia twigs (*utsugi*; Deutzia scabra). These little nails are heated in a frying pan immediately before they are used. This draws the remaining moisture out of them and makes them harder, and thus easier to drive in. Holes for the nails are pre-drilled slightly obliquely outwards, formerly with a drilling rod (*kiri*), now usually with a small hand drill. The proud sections of the wooden nails are cut off with a dovetail saw. These positions are moistened with a cotton cloth before the boxes are smoothed on all sides with the plane. To do this the little box is placed with the crafts-man's left hand against the stop on a work-surface made of a thick Keyaki plank; the plane has to be very well sharpened in order to take fine shavings of this specially soft material with a smoothing plane held in one hand. For higher-quality products the lid is planed to a slight curve and the edges are bevelled. Finally the box is polished, until 1970 with field horsetail and now with 320 and 400 grade sandpaper. The Kiri wood acquires the desired slight silky sheen from being rubbed down with a little cotton bag filled with a handful of *ibota-no-hana*.

Originally the Kôsaka workshop specialized entirely in boxes, but as demand has halved over the past 30 years they have diversified and make some items for the tea ceremony (*sadô-gu*), like little shelves and trays, for example.[3] Almost all the products in both groups are commissioned by five tea utensil wholesalers. It is tea scholars and the small but constant demand from museums and collectors that keep the box-maker's profession alive. One of the three employees will take the workshop over and continue to run it.

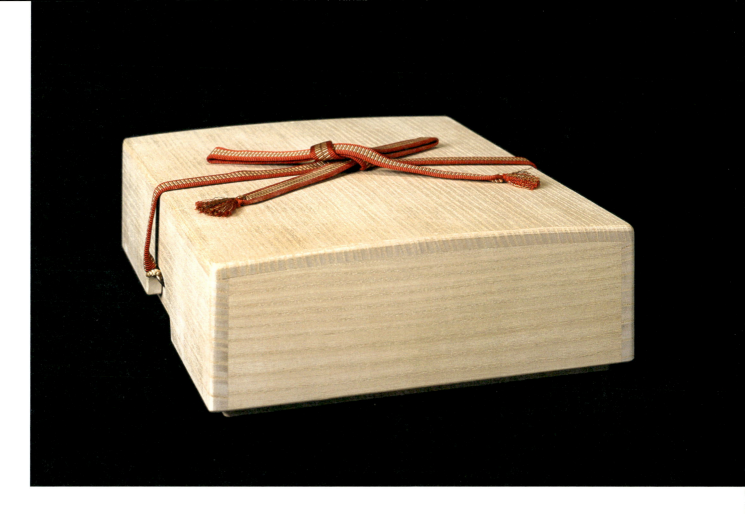

*Stringed box with
slightly curved
slip-on lid.*

Coopers' goods

Coopers make containers, usually round, consisting of a number of narrow upright staves and held together by hoops.[4] This trade did not emerge in Japan until the 14th or 15th century, but then spread rapidly, as shown by the name coopers' quarter in some towns. Japan did not have planes until the 14th century, which is probably why barrels and tubs did not appear in Japan until about 1500 years later than in Europe; a large plane or a cooper's jointer is needed after all to true up the edges of the staves. There are other things that are different from Europe as well. Japanese barrels are not made in the bulbous shape we are familiar with, they tend to be coniform here. In Europe, flat iron strips were used for the hoops from an early stage for large containers, while in Japan split bamboo cane is still used today. Barrels and tubs were made almost exclusively of coniferous wood, usually the relatively soft Japanese cedar (*Cryptomeria japonica*), until the manufacture of whisky barrels started in Japan in the late 19th century; here European models were followed completely and oak was used.

The region around the northern Japanese city of Ôdate can be considered the coopers' centre today, as it still has six workshops.[5] There are two particular reasons why this trade has survived here. The wooded region is known for its Japanese cedars, and because of its peripheral position and poor transport links it has not completely kept up with the speed of progress elsewhere.

Putting on the last hoop: the cooper turns the vessel bit by bit with his foot.

The Tanaka workshop

The Tanaka workshop is in a side street on the edge of the centre of Odate, next door to the family home. Father and son work in the spacious shop, both sitting on the floor. The tools hang in rank and file on the wall, or are stored in racks. The cooper's actual workplace is identified by a slice of a hardwood trunk about 80 cm in diameter and 40 cm thick, which is let into the floor, and protrudes the floor by a few centimetres. This block gives the cooper rigid support for his work when assembling the goods. A striking feature among the tools is a large cooper's jointer, almost 1.5 m long. This plane is turned over and held at an oblique angle by a board. The staves are run over it to true up their edges.

A thick hardwood slab is let into the boarded floor at each work-place. This serves as a firm base when assembling the barrels. Drawing knives, planes and templates hang on the walls.

About 30 m³ of locally grown Japanese cedar per year are used in this workshop. Tanaka first cuts the trunks to length with a chainsaw; they can be up to a metre thick, and can only be stored for two years as logs. Until ten years ago this job was done with a two-man cross-cut saw. He then splits the trunk sections with a straight cleaving knife (*owari-nata*) radially into at least eight segments, removing the immediate core area; these steps are called *ô-wari* or "coarse splitting". The necessary force comes from a large mallet with a round head secured with iron rings, wielded from a standing position. Now comes the "fine splitting" (*ko-wari*), in which the individual staves are split off from the segments. Here the craftsman has to be sure about what the vessel will be used for: if it is intended for storing liquids over a long period (*taru*), Tanaka will make the staves from timber with horizontal annual rings (*itame*), placing the cleaving knife parallel with the annual rings. If a container for rice or a ladle (*oke*) is being made, he places the cleaving knife at right angles to the annual rings, and the staves then have vertical rings (*masame*). The account below describes the cooper's technique for a rice wine barrel with a capacity of about 36 l, his most popular product.

The log sections are split radially into eight or more segments.

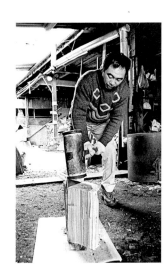

Rice wine barrel

As a rule the dark pink to brownish red core wood of the Japanese cedar is used for the staves of a rice wine barrel. But barrels with core wood on the inside and sapwood on the outside are also popular. Barrels like this, which are white outside and red inside are called *uchi-aka* ("red inside"). They used to be more expensive, as the material has to be chosen particularly carefully. Today Tanaka sells them for the same price, as he is pleased to have as little sapwood as possible left. Core wood is also used for the floor-bottom, and only the lid is made entirely of sapwood.

The cleaving knives used are slightly curved, which means a better yield, but also makes the next stage of the work easier, the coarse preparation of the staves with drawing knives. The cooper selects the appropriate cleaving knife from a wide range, according

The thickness of the logs is marked with the aid of the notched hammer handle. Two thick logs were cut off this segment first of all.

The thick logs are split in half to produce four sections at first, and then eight staves. The cleaving iron is curved like the barrel wall later.

to the diameter the barrel is intended to have. He places the segment on the hardwood block in front of him and holds it between his feet. Notches in his hammer handle help him to place the cleaving iron correctly. The staves are not now cut off one after the other, in fact two or three pieces four times the thickness of the barrel wall are cut, and then split down the middle. This means the cooper can be sure that the cleaving iron will not run off line, which can easily happen with thin pieces of wood. Another effect is that when splitting down the middle he can rely more on his eye.

The stave is clamped between stop and stomach protector so that it can be worked with the drawing knife.

Truing up the staves – the staves are run over the obliquely placed cooper's jointer.

Further work is now done on the staves with drawing knives, a straight one (*soto-sen*) for the edge and outside of the staves, and one with a concave curve (*uchi-sen*) for the inside. The cooper crouches with bent knees on a plank a good metre long, inclined slightly forward by pieces of boarding placed under it. A small piece of wood is fastened to the front of the plank. He holds the stave firmly between this stop and a wooden stomach protector held by his apron strings, and pulls the drawing knife towards him. He works on the two edges first, then the outside and finally the inside, with the curved drawing knife.

The staves, called *gawa* in Japanese (literally "sides"), are allowed to weather in the open air for two months. To make sure the weathering is even, the staves are piled into little frame-shaped towers. After that they dry in a shed for three months. Before the barrels and tubs are assembled the staves are smoothed with drawing knives (*gawa-zukuri*) and the edges trued up. The cooper pushes them over an inverted plane almost 1.5 m long (*shojiki-dai*), whose back end is set up obliquely with the aid of a board. The sole of the cooper's jointer is kept slightly concave (approx. 3 mm), so that the staves become slightly conical at the ends. This ensures that the barrel, whose staves are held together above all by the hoops at floor and lid level, does not leak halfway up. For a long time Tanaka had to use a wooden gauge (*kata*) to help him when truing up the staves, to check the angles, but now he does not need this any more. After a good 20 years' experience of his trade he works completely by instinct. His body has internalized all the positions, the sequence of work seems effortless and has acquired a rhythmic quality like the hiss of the plane.

The cooper starts the assembly stage by placing the staves upright (*gawa-tate*). Tanaka sits on a little bench in front of his working surface, which is let in the floor. He puts two setting hoops and some staves in place. The setting hoops are made of iron rods. He picks one up with his left hand and pushes it against his right knee. Then he gradually puts the staves in position with his right hand. There are different shapes. Most are almost parallel, but according to the size of the vessel four to six staves are strongly conical, and are called *ya-gawa*: an 18 l barrel has 4 to 5 such *ya-gawa*, a 36 l barrel usually 6 to 7 and for the largest, 72 l barrel, Tanaka uses 8 to 9 *ya-gawa*. The staves for a particular barrel are not put together precisely beforehand, Tanaka just counts the con-

Proud sections of the
outer wall are cut off
with the drawing
knife.

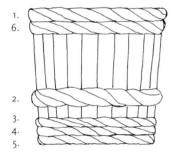

Barrel with hoops and the
sequence in which they
are fitted.

ical staves out. He usually has to swap some staves before he can close the vessel wall. He does this all by eye and does not measure anything.

The staves are held together first of all with two or three iron hoops, a round iron one at the top and one or two flat ones at the foot of the barrel. After the hoops have been pushed on, the outer barrel wall is smoothed with a very sharp drawing knife and the proud wood at the joints removed; Tanaka holds the barrel between his upper thighs while doing this.

Now the hoops called *taga* are pushed on to the body of the barrel one after the other. They come from strips of bamboo just under 6 m long, tapering from 24 to 6 mm in width. They should be used within two months of the felling of the bamboo, because unduly dry bamboo is more difficult to work and prone to cracking. The cooper makes his own hoops from the long strips and has to adapt them precisely to the size of the vessel. He starts with the cap hoop (*kuchi-wa*) at the top end of the barrel. The hoops are pushed on with the aid of a hoop driver made of hard maple wood (*shimegi*) with its end clad in sheet iron, and a heavy wooden mallet (*kizuchi*). After two powerful blows the cooper turns the barrel a few degrees with his feet. This rhythmic flow of double blows is the acoustic trademark of the craft. The cap hoop is followed by the "first body hoop" (*dowa no ichiban*), which is placed below it.

Before more hoops are added, the inner wall of the vessel is smoothed with a round plane (*uchi-ganna*). The sole of this plane is very slightly curved longitudinally. A smooth, perfectly aligned wall is essential for fitting the bottom. Tanaka works on the bottom (*soko or soko-ita*), which today is first cut roughly with a band-saw, with the drawing knife, after moistening the edge of the board, as wet wood is much easier to work with. Tanaka then places the bottom upright, using the mallet he has put on the work surface as a stop. First he bevels the under-side heavily and smoothes the edge. The prepared bottom is then put into the barrel from the top and beaten temporarily into position with a slightly conical wooden bat about 70 cm long (*soko-otoshi*, literally "beat down bottom") made of hard Keyaki or maple wood. The bottom is tested to see that it fits precisely, and then taken out again. The groove for the bottom head is cut with a hooked blade fastened to a long handle. It is only about 1.5 mm deep and round in cross-section. Finally the edges of the bottom are moistened again, it is put in from the

The hoops are made from long strips of split bamboo.

Pushing the first hoop on with a hardwood hoop driver.

Smoothing the inner wall with a round plane.

The bottom is driven in from above with a wooden bat.

top, batted into position and the "body hoop" is pushed higher up. The two iron hoops remain in place.

Then comes the third hoop, known as the "bottom holder" (*soko-mochi*). It is mounted from a sitting position. The cooper holds the barrel with both feet in such a way that his big toes can grip the barrel wall. Tanaka hits the hoop driver twice with the mallet and then turns the vessel slightly clockwise with his left hand; he takes his feet off the barrel wall for these short moments during which he is turning it. His feet are so "nimble" only because he is wearing the traditional shoes known as *zori*. They are very flexible,

Cutting the groove
for the barrel lid; the
upright barrel is
pushed against the
left leg to do this.

Levering a bamboo
hoop.

and made of a stout cotton material; the big toe is covered separately, and can be spread out, which makes it possible to grip the barrel wall. After the hoops have been fitted the protruding ends of the bamboo strips are cut off with a long flat steel bar with a short oblique cutting edge ground into its end (*motogiri*). Tanaka places the blade carefully and taps lightly with the mallet.

Now the iron hoop at the bottom can be taken off and the second bottom hoop (*soko-mochi no niban*) fitted. Tanaka holds the bottom hoop at the end of the inverted barrel with his feet and levers the hoop on with the blunt end of the *motogiri*, which tapers to a cone. The critical moment comes when the fifth hoop is fitted to the bottom end of the barrel. The coopers call this the hoop of tears (*naki-taga*), as it takes great effort and dexterity to fit it because so little room remains. It has to be levered on like the lower of the two bottom hoops.

Finally the end grain at the foot of the barrel is smoothed with a straight drawing knife (*koguchi-kiri*). The edge is also slightly bevelled outwards and then milled on the inside. This prevents tear-out at the end-grain.

To press the staves against the bottom of the barrel the cooper hammers as hard as he can against the hoop set level with the bottom. This high pressure causes the soft Japanese cedar wood stave to distort slightly, any cavities are closed and the barrel is made tighter at this crucial point. The end grain of the staves is slightly bevelled and the outer edge lightly milled at the upper edge of the barrel.

At last the top head, or lid, which is made of several sapwood boards, can be put into position. Its edge is moistened, the cooper stands it on its end between his feet and works on it with the drawing knife, as he did on the bottom. The lid had already been cut out at an angle of about 30° on the band-saw by setting the table diagonally. The lower edge is milled first, before the bevel is smoothed. Tanaka turns the lid as needed so that the knife is not running "against the wood", and to prevent tear-off. A groove is now cut at the top of the barrel to take the lid. Here Tanaka uses his left foot as a stop, and puts his right foot on the barrel wall. The long handle and the leverage it produces make it possible to work apparently effortlessly. The edge of the lid is moistened again so that it does not slip when being put in place. Now the cap hoop can be driven in as far as the top edge of the barrel. Before this is done the remaining round iron

setting hoop is taken off the top edge of the barrel. Tanaka knocks it off with a single well-aimed blow of the mallet. Here too the bamboo hoop is hammered with the mallet from the side, to marry the staves to the lid.

The last hoop, a second cap hoop (*kuchiwa no niban*), is fitted. There is no gap between it and the first cap hoop, they are in contact. Before the barrel goes into storage it is tested above all for seal tightness; Tanaka introduces compressed air through the lid aperture. About a tenth of the barrels produced are not quite airtight. Usually hammering the hoops is sufficient to correct this. If there are flaws in the wood a little hole is drilled and then stopped with a coniform piece of wood.

Coopers' goods: types and uses

Tanaka sells to four wholesalers and one of the four remaining local rice wineries. His best customer is a wholesaler in Shiga near Kyoto, who takes roughly a third of his production. The wholesalers usually order five or six barrels, in the past it could have been 30 or even 50. Unlike the barrels, Tanaka usually sells his little tubs directly; he makes them in series of 15. Vessels for boiled rice (*ohitsu*) sell particularly well; every household used to have one until the sudden introduction of electric rice cookers, but they are now luxury items in traditional kitchens, and the same applies to vessels for pickling vegetables (*tsukemono-oke*). He also produces smaller scoops in bigger runs; these are the kind that used to be found in Japanese baths. Such scoops, usually made of Hiba wood, which is particularly water resistant, have been almost completely forced out by cheap plastic substitutes, though their popularity has increased again.

The workshop produces about 200 barrels per year with a capacity of 4 *to*, 350 barrels in each of the sizes 2 *to* or 1 *to*, and 50 little barrels (*nishô*). To (18.039 l) and *shô* (1.8 l) are old liquid measures that were actually replaced over a 100 years ago with Western measurement by litres, but they have survived in this traditional craft. Incidentally old measurements have also shown great tenacity in the timber trade, on building sites or in tool manufacture.

The father has put the cooper's jointer against his upper thigh, and by his right hand he has placed the little staves for the scoops used in the Japanese bath.

The rice wine barrels, which now represent about 80 % of the workshop's turnover, are traditionally used on solemn occasions like topping-out ceremonies, opening ceremonies or also for sporting victories; they often turn up during election campaigns as well. The wine is kept in them for only a few days, until it has taken on the taste of the wood. Before the festive event the rice wine merchant loosens the cap hoop of the barrel a little, otherwise it would scarcely be possible to knock the lid in with the mallet. Once the lid has been knocked in, to the accompaniment of applause and congratulations, and the wine consumed, these containers have actually done their job. Until the post-war period it was customary to pickle vegetables in the barrels afterwards, but such secondary uses are rarely seen nowadays.

The 40-year-old Tanaka is not only practising a trade that has become very rare, he also plays a special part among the few remaining coopers in two respects. A lot of coopers have rationalized their production considerably. The staves are cut with a mechanical saw, not even trued up, and the barrel is turned on a lathe. Tanaka works just as his predecessors did a 100 or 200 years ago, apart from the use of planing machine and band-saw for bottoms and lids. Most coopers make nothing but rice wine barrels nowadays, the only product for which there is a reasonably steady demand. Tanaka, whose workshop originally specialized in bath-tubs, still builds around four of these large items per year, and also tubs for pickled vegetables, rice, and even containers for ritual purposes. There is probably no other cooper in Japan who practises the cooper's craft in this unbroken way, and has a command of such a wide repertoire of forms.

Store with semi-finished parts used to make containers for rice and pickled vegetables, ladles and scoops.

Turned wooden cores for lacquered bowls

Lacquered goods are one of the best-known and internationally acknowledged branches of the applied arts in Japan.[6] Apart from the early dry lacquer figures in 8th to 12th century temples and free art-works they have a wooden core as a rule. They were made using split labour from an early stage. There were craftsmen who concentrated completely in making the wooden core, *kiji* in Japanese (from the two characters for "tree" and "ground"). Four different approaches can be distinguished, according to technique. If the wooden core is made of thin boards, as for caskets or certain kinds of tray, the technique is called *sashimono-kiji*. Wooden cores with bodies made of thin bent wood are called *magemono-kiji*, and carved cores as for certain bowl forms are called *kuri-mono-kiji*. But the wooden core is most frequently turned, which is called *hikimono-kiji*. There are and were particular workshops for all these techniques.

Most of the workshops making turned wooden cores have specialized in soup-bowls.[7] There is even a special name for this profession, *wan-kiji*. As soup is practically obligatory in a Japanese meal, it is scarcely possible to imagine Japanese tables without these bowls, which are usually lidded. Wood finished with natural lacquer is not just preferred for aesthetic reasons; wood is less conductive than ceramics or porcelain, which makes it pleasanter to handle.

Aizu-Wakamatsu in northern Japan or Kainan near Wakayama are among the places that are home to the lacquered goods industry, but the little town of Wajima on the remote Nôto peninsula on the Sea of Japan is considered to be the undisputed centre of this craft. A 100 years ago there were still 300 bowl-turners here; the figure has now gone down to 25, working in twelve little workshops. This sudden reduction reflects several developments: changed eating habits, the manufacture of plastic substitutes, produc-

Wooden cores for soup bowls – precise manufacture in small series.

138

tion on computer-controlled lathes and the import of ready-primed semi-finished goods from neighbouring China.

The Wajima bowl-turners work as manufacturers of semi-finished goods right in the shadow of the famous lacquer workshops. But lack of recognition does not seem to have put the younger generation off: the average age of the bowl-turners is only just over 40. So there is no immediate concern about this profession's survival.

Material

Locally available deciduous timber is always used for the turned wooden cores. Hard Keyaki wood is preferred in Wajima. It occurs frequently in the region, is easy to turn, lasts well and takes natural lacquer outstandingly. The trees, which should be at least a good 100 years old, are felled in the winter months. There is no felling during the *hassen*, a twelve-day phase occurring six times a year according to the old Japanese calendar. It is said that timber from trees felled in this particularly wet period gives wood that does not last well and also twists a great deal. The Keyaki trunks are left lying for two to three years before being cut into pieces known as *kata* at the beginning of autumn. Until the seventies these pieces were cut out of the trunk with axes and then coarsely finished with a machete and a short-shafted adze. Today the trunks are cut into thick planks on a band-saw, or from the trunk in discs. The blanks, which are cut from the planks or discs with a band-saw are turned cylindrically and hollowed out on the inside.

The blanks are dried in an unusual and elaborate way. They are piled up carefully on the top level of small wooden sheds, and smoked from below over a period of a good four weeks. For this "smoke-drying" (*kunei-kansô*), which has the effect of a kiln, the available shavings are charred in an iron bowl. The blanks take on a dark brown to black colour because of the soot. Until ten years ago the turners did their own smoking, today it is done mainly by the blank manufacturer (*kata-ya*). Near Wajima, in Minami near Ozawa and in Yamashita near Sôryô there are two one-man businesses that make and smoke these blanks.

The circumference of the bowl is marked on the blank with a template.

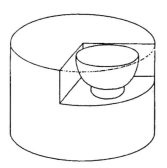

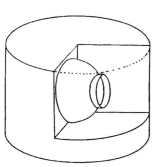

There are basically two methods for selecting timber: the vessel can be turned using "vertical wood" (*tategi-dori*; the fibres lie parallel to the axis of the vessel) or of "horizontal wood" (*yokogi-dori*, with the fibres at right angles to the axis of the vessel). Vessels with vertical fibres, in other words made of long-grain timber, are relatively delicate, but distort less and the blanks dry out within a month. Vessels in cross-grain timber are more stable but distort more easily and the blank should dry out for six more months after smoking. Thin-walled vessels tend to be made of cross-grain timber, but today the majority is made of long-grain timbers.

Timber selection possibilities – bowl in long-grain (tategi) and cross-grain (yokogi) timber.

Tools

The turners are probably the only wood-working profession in Japan who still forge their own tools. This is all the more surprising because Japan has a great forging tradition and there are large numbers of highly specialized tool-smiths, especially for woodwork, making planes, for example, chisels, carving knives, hammers or saw-blades. About every three months the turner spends a day forging about 20 turning tools, about four or five to keep in stock of each type of turning chisel.

The Kanchô workshop

The Kanchô workshop will be described here as an example of bowl-turning in the lacquer capital. It is only a block away from the port and is directed in the fourth generation by Kanchô Shôzô, who was born in 1961; he started work for his father immediately after leaving senior school. His first name, Shôzô, which is written with the two characters for "right" and "making", seems almost like a motto for the high-quality work of this trade.

The single-storey workshop with a storeroom for semi-finished products was built behind the family home in 1919; it almost resembles a dungeon with its semi-darkness and sawdust everywhere. But this has obviously not affected the turner's spirits as he sits on the wooden floorboards at his electrically powered lathe cross-legged, with his peaked cap on back to front. It is a facing lathe with bowl-mounted tailstock, which has a chuck screwed on to its top. So the workpiece is not fixed between the head- and tailstocks, but to different chucks, and always "free-turned". In front of the chuck there is a

Turning the interior shape.

Turning the outer wall.

section cut out in the floor of about 40 x 40 cm, making it possible to work with large pieces, and making room for the shavings that will be produced in great quantities. The turner can select one of three speeds by moving the drive belt, according to the size of the workpiece. The tailstock turns clockwise.

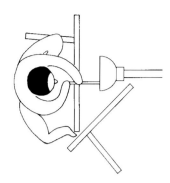

Kanchô works only with indigenous deciduous timber, about 80% Keyaki, the remainder is wood from the Gingko tree, called Itchô. The latter lasts very well and is easy to work, but sometimes has very large resin pores (*dôkan*). To make centring easier, the turner places a turned wooden template (*jôgi*) on the sooty blank and runs a pencil round it. Then he places the blank on the spiked chuck on the tailhead and hammers it into place with a few sharp blows from a wooden bat about 50 cm long (*shumoku-bô*); this process is called *nomi-date*, literally "setting up the turning iron". He turns the blank a little by hand, then switches the lathe on, to make certain that the blank is properly centred. Now he puts the support for his turning chisels in the correct position; this is a board, about 50 cm long and 2 cm thick, placed upright, with a batten tenoned into its back at an angle of about 40°. His left knee is on this batten, and his right foot is braced against the right-hand end of the support. This support is called *kanna-makura*, a "cushion for the turning iron". Before work begins, the turning chisels are sharpened with an 800x waterstone.

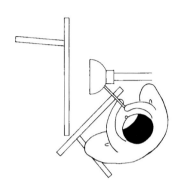

First the basic shape is created, *ara-kezuri* ("coarse shaving removal"). The turner sits on the extended axis of the tailstock and turns the inside of the bowl with the chisel, which is semi-circular at the tip (*nakaguri*). He checks the depth of the vessel with a depth gauge (*uchimi-bô*). This is a an iron bar about 25 cm long, which has a pin set in the middle at a right angle, and this can be pushed out to indicate the appropriate depth and fixed with a screw.

*The turner sits along the tailstock axis or parallel to it.
He changes his position repeatedly.*

Checking the vessel profile with a template.

Putting an anti-wobble wooden ring in position.

The turner now changes his position very rapidly to work on the outer wall. He sits parallel with the tailstock, where a second support for the chisels is in place. Once the outer wall has been coarsely worked he checks the height of the bowl with the length gauge; this looks like a sliding calliper, and is fitted to the lathe parallel with the tailstock. Finally the foot of the bowl is turned. The basic shape has now been created, and it is carefully removed and dried for about two months in the upper section of the storeroom, known as *ama* (sky). This pause is followed by fine work on the vessel (*shiage*). Kanchô starts on the outer wall, taking off just a little material with the *metori-ganna*. The lathe is stopped several times so that the vessel's profile can be checked with a template (*kai-gata*). There are whole boxfuls of them in Kanchô's workshop, showing

141

Turned wooden cores for lacquered bowls

the profile of the bowl (*oya*, literally parents) and the appropriate lid (*futa*). Until a few years ago they were made of strips of wood about 3 mm thick, but now Kanchô uses thin white acrylic sheets, which are easier to work. Sometimes a whistling sound starts up while turning, caused by an imbalance in the thin-walled vessel. Kanchô then stops the lathe, picks up a handful of shavings with his left hand and pushes them into the inside of the bowl. Then he starts the lathe again, the shavings are thrown against the wall of the vessel by centrifugal force, and stabilize it.

Smoothing the inner wall; the edge is secured by the wooden ring.

Checking the depth of the vessel, important for stacking the bowls.

Securing the bowl to turn the foot.

Turning the under-side of the foot with a bottom steel.

The turner changes his position again so that he can smooth the inside of the bowl (*naka no shiage*), now sitting on the extended axis of the tailstock again. The edge of the vessel (*uwabuchi*) is slightly bevelled, the lathe is stopped and a wooden ring (*ki no wa*) is placed on top. Once the thin vessel wall has been secured against wobble in this way by the ring, which is turned in Keyaki wood and about 1 cm wide, the inner wall can be turned with the *nakakuri*. The lathe is stopped again and the depths of the bowl checked once more. Irregularities would scarcely be noticed in actual use, but they show up when the bowls are stacked in their storage boxes. Finally the turner smoothes the outer wall with 120x sandpaper, so that the primer will take better.

The turner sits parallel with the tailstock again to cut off the conical foot with a *tsukiri* (parting chisel). He angles the turning chisel slightly inwards. He knows the precise moment at which the vessel will detach itself, and catches it safely with his left hand.

The relatively high foot of the bowls is turned. Its walls are wafer-thin and called *ito-wa* ("thread ring"). First the chuck has to be changed. The spiked chuck with its ring of steel points is replaced with a chuck that has a round wooden disc mounted on it (*uragaeshi-dai*). It has a turned groove to receive the lip of the bowl. A wooden ring (*urakaeshi-wa*) is placed on the bowl foot to hold the vessel. It is extended like a bar in the middle. A string is wrapped around this extension and a wooden bar (*rikimi-dake*), which is attached to the chuck through a drill hole. Secured in this way, the foot can now be turned. A floor chisel (*nakakuri*) is used first, before the foot is smoothed off with the metori. Finally the foot end is smoothed with 120x paper.

Work on the bowl and the lid is the same to a large extent; the lid is slightly smaller in diameter, and the wall slopes considerably less, but the basic shape is the same.

Working conditions

Kanchô produces about 30 bowls a day in this way. They represent about 80 % of the workshop's output, the rest are trays. Kanchô turns about 1800 bowls with lids per year. He is paid about 3000 yen per bowl. He himself pays about 750 yen for a ready-smoked blank of average quality, 450 for the bottom section and 300 yen for the somewhat lower blank for the lid. Blanks of particularly fine-grained wood cost up to double. Kanchô's work brings him an annual income of about four million yen.

All Kanchô's commissions come from the town or the immediate vicinity, a good 90 % from large lacquer workshops, the rest from artists. The formal repertoire has scarcely changed over the generations. But unlike former times, dimensions have been fixed individually in the last ten years, and the old standard measurements for vessels are followed increasingly infrequently. As well as the usual smooth-walled vessels, bowls decorated with fluting (*sensuji*), cords (*maru-itome*) or a large number of fillets (*hi-itome*) are produced. They are usually finished just with clear Urushi varnish and thus require immaculate material and almost three times the amount of time.

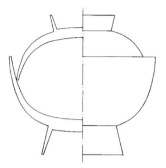

View and section of the soup bowl illustrated here.

Samples of the major processing stages.

Hoshi with a stitched
box body in Japanese
cedar.

Chip boxes

One of the first wooden containers made by man was the chip box. These extremely light vessels are usually cylindrical or oval, with a body made by bending bark or thin board.[8] They were found in many different forms in Japan, where they are called *mage-mono* ("bent thing"). Usually wood from conifers like cedar and cypress was used, as it grows straight and is easy to split. Until the late Middle Ages, even large vessels were made in this technique, and used for storing and transporting rice, vegetables and even drinks, for washing and also for storing valuables.[9] When containers made of staves appeared in Japan as well around the year 1400, chip boxes became less important, and the range subsequently reduced to small objects like boxes and ladles. In Japan making chip boxes became a profession in its own right in the 18th century at the latest. Until the post-war years chip box makers (*mage-shi*) were to be found in many places; their principal products included boxes for provisions called *bento-bako*, which are comparable with our stacked portable food containers. Then in the fifties chip boxes were almost completely forced out in Japan by industrially produced aluminium, then later plastic, containers. After that the profession survived only in a few little niches.[10] A small workshop in northern Japan serves as an example of this.

The Hoshi workshop is in the mountain village of Hinoemata-mura, literally "cypress branch fork", in a wooded, snowy region in the Fukushima Prefecture. Making chip boxes has a long tradition here. Until the fifties, many men went into the woods to split by hand cedar trunks that had been cut down to lengths of a good 70 cm. The chips were then smoothed with the drawing knife and then rolled. "Nests" of this kind, made up of ten bent chips, were tied together with straw, taken to the village and sold to a dealer as semi-finished goods. They were made into boxes and sieves in the provincial town, and mainly finished with clear varnish. Containers for personal use were the only ones made in the village. In about 1960, both branches of the trade practically died out. But Hoshi and a few others in the village took the craft up again in the mid seventies, when traditional wooden products were once more enjoying greater popularity. He is now 74 years old, and still makes a good 1000 boxes each year.

His little workshop, which has firewood stacked against the outside walls, measures just three by three metres. The wooden hut may look primitive at first glance, but it is very clean and neat inside. The floor is covered with straw mats, and of course shoes are left at the entrance. In the corner on top of a little stove is a tub of water, patterns and tem-

The body-boards, about 2 mm thick, are made supple in a bath of hot water.

plates hang on the walls, and thin boards, his raw material, dry as they are hanging from the ceiling. These boards, split earlier, have been cut with a band-saw, thicknessed and even finished with an automatic finishing plane in the village's own sawmill for some decades. Apart from this preliminary work, scarcely any other aspects of the process have changed. The chip boxes are the same as the earliest surviving examples from the 8th century in almost every detail.[11]

First Hoshi planes the ends of the ready-cut 1.5 to 2.5 mm thick bodies obliquely, so that the chips taper. This tapering ensures an almost continuous transition of the later overlap on the body. Then the bodies are placed in a bath of water at a temperature of about 80° Celsius for about ten minutes, which makes them astonishingly pliable. Hoshi takes the little boards out and puts them on a strip of sailcloth whose end is fastened to a cylindrical wooden block (*koro*). He now rolls this block slowly over the body, the chip is wrapped up between the sailcloth and the block and thus bent. This process has to be repeated several times if necessary, until the body is bent to the right extent. The bent chip is then fixed by attaching a small clamp – called *hasami* in Japanese. This clamp is as primitive as it is effective: two battens have string tied round them at one end, at the other end they taper slightly conically. Once the battens have been pushed on to the body of the box a wooden ring or clamp is positioned at the conical end, thus fixing the body between the battens.

The bodies, still steaming, are wrapped between sailcloth and a wooden cylinder.

145

After the bent bodies have been dried for a few days, glue is applied to the butt joint. A paste made of crushed rice grains or hide glue was formerly used at this stage, today it is usually white glue. Then the wooden clamp is replaced, fine slits are cut with a knife and the overlap is "stitched" with narrow strips of dark red cherry bark. This seam was originally merely a functional way of joining the ends of the body permanently. Finally the somewhat thicker bottom can be fitted and fixed with small wooden or bamboo pins.

Putting on the wooden clamp.

The bodies have to dry out over night.

Working on the strips of cherry bark with a knife. A bare foot is used as a rest.

The overlap is sewn with bark. First Hoshi has to make fine slits with a thin knife. He has marked the spacing of the slits on the wooden clamp.

146

*Four types of boxes
for provisions —
differently from modern
arts and crafts work,
the white sapwood of
the Japanese cedar is
used as well.*

Hoshi with his wife.

Stacks of thin, bent boards used to make bodies and hoops are stored on the top floor of the workshop.

Sieves and steamers

Sieves and steamers, as well as containers, are objects that are made by bending thin boards.[12] The Japanese distinguish between two types of sieves: in the case of the *furui* type the woven metal or silk is on the bottom; thus food, sand or other items are shaken through the sieve from above, as in the West. A second type is called *ura-goshi*, "inverted sieving"; here a horsehair mesh is fitted on the top and food like boiled eggs or the chestnuts often used for sweet dishes are forced through the sieve with a spatula and thus finely puréed. Both utensils are still used in traditional Japanese cookery. For steamers (*seirô*) a mesh of bamboo strips is fitted at the bottom.

There was never a particular centre for this craft, and the few remaining workshops are spread all over the country. There are records of a sieve-makers' association in the early 19th century in Yamada, a village on the Sea of Japan (on the side of the main Japanese island, Honshû, facing Korea) that still lives mainly on fishing. Incidentally, the main reason for this association was anxiety about over-production and a price collapse; so the craftsmen were only allowed to train apprentices with the agreement of their colleagues and under certain conditions.[13] After the Second World War there were still more than ten families who made sieves as their principal occupation or as a sideline, today only the Adachi workshop remains, which can look back on a tradition going back

ten generations. Adachi's workshop is opposite his home on the land side of the coast road, which in Yamada is fringed with carefully well-kept timber-clad houses.

Material

Until the late 19th century local Japanese yew (*Kusa-maki*) was used for sieves and steamers, but as stocks dwindled it was replaced by high-quality cypress wood from Kiso in the Nagano Prefecture. Since 1930, close-grained pine wood from the northern island of Hokkaido has been used. This kind of wood is called Ezo-Matsu, after the old name for the island and its original inhabitants.

For over four decades, Adachi obtained the little, thin, bent boards that make the starting-point for the bodies from Fujimoto in Kitami on Hokkaidô, who specialized in these semi-finished goods. He split the pine trunks radially, so as to have boards with vertical annual rings. They were then worked with the drawing knife and after two or three months' natural drying were boiled for one to two hours, thus for a relatively long time; this made them supple, and also the resin was drawn out of them. Once they came out of the water bath they were bent immediately, and placed in a prepared wooden hoop stitched with cherry bark. This produced, according to thickness, nests of 8-15 boards placed one inside the other, called *magi-wa* (bent hoop). Fujimoto delivered this base material in different widths: the small hoops used for the high substructure of the sieves are called *katsura*, and the wide ones used for the walls of steamers are called *oya-wa*

Taking out a bent pine board.

1.

2.

3.

4.

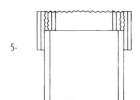

5.

"parent hoop". When his supplier gave up work on grounds of age in 1988 Adachi bought up all his stock without further ado. He is still using them, but he will have to look for a new source in five to eight years.

Manufacture

Something that looks at first glance like a body made from a single board is in fact a structure made of six thin wooden hoops set precisely one inside the other and joined to each other. A start is made with the two outer hoops, called *koshi-wa* (hip hoops) or *katsura*. The radial splitting gives each of the thin boards a slightly wedge-shaped cross-section. First shavings are taken off the inside of the hoops with a short knife (*ko-gatana*), to make their cross-section approximately rectangular. Adachi gradually turns the upright hoop on his workbench to do this. He then planes one edge of the hoop, which is still held upright, and reduces it to the required width with a cutting gauge. This cutting gauge with its protruding knife is run along the inside and the outside. This makes a deep cut in the relatively soft wood, so that the hoop and the excess material can easily be pulled apart. The two hoops are bevelled with the knife at the overlap (*awaseme wo kezuru*) and then stitched with a strip of cherry bark about 8 mm wide, first the inner and then the outer hoop. Cherry bark has been used from the earliest times to make bent containers, as it shrinks when wet and thus holds the body of the sieve together; its woven section is moistened and tensioned with use. Adachi applies glue to the inside of the outer hoop and taps the inner hoop into place. Rice glue was typically used, now it is replaced by white glue.

Three half-width inner hoops are set into the two outside hoops, so-called little hoops (*ko-wa*) or tightening hoops (*hari-wa*). They too are cut from wider boards with the cutting gauge, and their insides are trimmed with a knife before they are cut to length with a dovetail saw. Unlike the outer hoops, the narrow inner hoops do not overlap, they are butt-jointed. To make sure the joint is entirely fast and to ensure greater durability for this utensil, which often gets wet in use or when being cleaned, the top wood of the first and second inner hoop is charred with a lighted match.

*Putting on the horse-
hair mesh, which has
been soaked in water,
and placing the first
inner hoop in
position.*

The next step is to stretch a mesh of horsehair from a special weaving mill, now usually imported from China, over the hoops (*uma no ke-ami wo haru*). It is first plunged
into water; this makes it more supple. The mesh is cut into a square with scissors,
placed over the two outer hoops and the first narrow hoop is pushed into place inside.
Adachi pulls on the mesh from the sides to tighten it. Then he taps the inner hoop with
a wooden mallet until it is flush. The mesh is then turned inwards over the first inner

*The mesh is pulled
tight.*

*Putting in the second
and third inner
hoops.*

hoop and the second narrow hoop is immediately put in position. A third hoop follows,
for reinforcement only, before the excess from the horsehair mesh can be cut off with
scissors. The three inner hoops, fixed together by six little nails, are finally driven down
the outer hoop with hoop driver and mallet, until the mesh is flush with the upper edge
of the outer hoop.

The steps described so far also apply to the manufacture of ordinary sieves. But here
metal or silk meshes are used. But the "inverted" sieves are mounted on a high base
(*dai*) so that there is room for the sieved food under the mesh. This base is bent from
a wide board and held together at the overlap by three parallel seams made from strips
of cherry bark. It is placed inside the inverted sieve (*dai wo hameru*) and then stitched
to this in three to four places (*dai wo tomeru*). The slits for the strips of bark, which are
scraped and cut to size with scissors, are individually cut in advance by hand. The edge

*The edge is smoothed
with the plane.*

Sieves and steamers

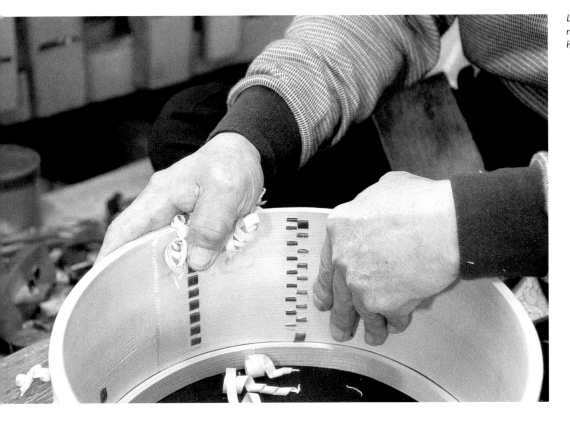

of the base is finally smoothed with the plane and rubbed down with a handful of shavings (*kanakuzu de kosuru*). This mills the edge slightly.

Every product from the workshop is marked by branding: the mark contains the name Adachi in an oval in phonetic transcription characters. The branding is done by Adachi's wife, who also looks after the store where all the different meshes are kept. About 60 % of Adachi's production goes directly to confectioners and traditional Japanese restaurants. The sieves are offered with a wide bamboo bark mesh, as well as horsehair. Sieves for craft and industry, usually with an iron, stainless steel or brass wire mesh, as used in iron foundries for moulding sand, for example, go to a wholesaler in Nagaoka. Since 1980, Adachi also makes containers from bent Japanese cedar, like small jugs and beakers for rice wine. They are sold by mail order, but account for less than 20 % of the turnover.

Adachi with a
completed sieve.

Kaba-zaiku – vessels made of cherry bark

Objects are still made from the barks of some trees in Japan, right down to the present day. The roofs of places of worship are often covered with cypress-bark shingles, and in northern Japan the bark of the crimson glory vine (*yama-budô*, Vitis coignetiae) is cut into strips for basket-weaving. The bark of the wild cherry (*yama-zakura*, Prunus sargentii) has been used in Japan since the 8th century at the latest. For example, a paintbrush has survived in the Shôsôin in Nara with cherry bark glued on to its handle. The handles and sheaths of machetes used for forestry work were also wrapped in cherry bark. One of the earliest uses, still practised today, is for chip boxes and sieves, whose bodies are stitched together with narrow strips of cherry bark (*tsuzuri-me*).

Kaba-zaiku, the use of cherry bark for arts and crafts, has a tradition that goes back a good 200 years.[14] The bark is prized for its natural markings, and made into precise containers. The origins are in the Aikawa region, in the north of the main island of Honshû. It is said that Samurai used to improve their income by working at home. The main products were containers for seals and medicine at first (*inrô*), and with the proliferation of tobacco consumption, introduced to Japan by the Portuguese, they shifted to *dôran*; these are two containers held together by a cord and dangling at a belt, a thin cylinder for a small pipe and a shallow container for tobacco, usually rounded on both sides (*tabako-ire*). So these were very small containers, as the relatively smooth bark of young cherry trees was being used, between whose striking stripes it was not possible to cut strips more than 15 cm wide. This technique is called *kata-mono* (mouldthing), because a wooden mould is used first of all to make a core of thin wooden strips; then cherry bark was stuck on to them inside and out. As the containers became more popular, and as immaculate bark from young trees became scarcer, in the early

The father at his workplace – his work surface is a tree-stump, next to it is a charcoal pan and the metal charcoal bucket, scrapers and abrasive devices hang on the walls.

154

20th century craftsmen went over to using bark from older cherry trees as well, detached in large pieces (*dai-han-gawa* or *omote-gawa*). This has a surface with some relief, and particularly striking markings. The product range consequently became larger, and little boxes made of thin boards for writing utensils (*tsuzuri-bako*) or trays also had bark stuck on to them. This technique is called *kiji-mono* (woodcore things), to distinguish it.

Working with cherry bark developed into a local industry in the little town of Kakunodate in the south of the Akita Prefecture. Machines for truing and polishing the bark were developed as early as in the thirties. Production was increased above all by replacing the elaborate wooden core with sheet metal, glass and since the fifties increasingly frequently with plastic. The number of workshops, which reached its maximum in the sixties at 120, had gone down to 60 by the year 2000. It is the recession and cheap imported goods from China that threatens the craft.

The Ogasawara workshop

In the 19th century Tashiro-machi, not far from Odate, became another centre for processing cherry bark; once there were ten workshops, now only Ogasawara's remains. Unlike most of the businesses in Kakunodate, this shop still works by hand, and the containers all have the traditional wooden core.

The Ogasawara workshop, a simple two-storey timber building clad with corrugated steel, is directly opposite his home. The father sits cross-legged in front of a tree-stump about 25 cm high and 30 cm in diameter. Next to it is a large round cast-iron charcoal pan (*hibachi*) where the irons are heated up and the bone glue is kept liquid. Some scrapers, knives and bundles of field horsetail hang on the walls.

Material

For cylindrical containers, the core, which subsequently has cherry bark glued to it, is built up from several layers of thin wooden strips. They are called *kyôgi*, Sutra wood, as Buddhist texts used to be written on wooden strips like this. Originally the container cores were made of bark as well, in fact from the bark of the Tsuki-no-ki tree. Today a veneer of Hono-ki, Sawa-gurumi and Itaya-kaede is used in the main. This shortens the manufacturing process, but does not ensure the same flexibility as strips of bark. The 40-year-old son demonstrates playfully: the bark strip bends almost like rubber, the veneer strip breaks after a few movements.

The bark is smoothed with a scraper.

The workshops in Kakunodate get their cherry bark from a few dealers in the region and from Yoshino in the Nara Prefecture, where peasants and forestry workers collect the bark as an additional source of income. The bundled bark costs about 1000 yen per kilogram. But Ogasawara rarely buys bark in, since 1990 he has usually harvested it himself between mid July and September (*kawa wo hagu*); this means he can be sure

The bark is bevelled at the overlap point.

Bone glue is applied to the overlap with a little wooden stick.

The strip of bark is carefully put together.

that the most attractive bark is not destroyed through lack of expertise. The strips of bark are about 15 cm wide and 40 cm long, and are dried naturally in bundles for at least two years. Once dried, the material keeps for an almost unlimited period, so the workshop has bark stocks that are over a 100 years old.

The iron is passed over a piece of field horsetail and drawn over the glued edge.

Manufacture

Cylindrical containers, the usual type for cherry bark work, and at the same time the speciality of the Ogasawara workshop, are made with the help of wooden moulds, usually in two parts. Ogasawara turns them himself from a kind of maple called Itaya-Kaede. Before being worked on the lathe the two halves are temporarily attached to each other at a few points with bone glue; they can easily be taken apart by soaking in hot water. The workshop has about 200 of these moulds.

First the core is made, a procedure called *shitaji*. Ogasawara shows us the steps taking a small tobacco box (*tabako-ire*) as an example, of the kind popularly worn on a belt

A second ring is pushed inside.

After glue has been applied on both sides, the cherry bark can be laid over the core and ironed on.

in the Edo period. The core is made of three thin strips of bark, 0.7 to 1 mm, from the Tsuki-no-ki tree. Ogasawara moistens the bark strips with an old toothbrush to make them supple. Then he cuts them roughly to size with scissors, smoothes them with the scraper, which is about 8 cm wide and called *tate-ganna* (vertical plane), then tapers them at the overlap point. The piece of bark for the outer layer of the core is now cut to its final dimensions with a steel square and bone glue is applied to the overlap. Ogasawara draws the iron that has been heated on the charcoal over a piece of field horsetail on his block of wood to ensure that no bits of dirt are sticking to the bottom. These irons, used in about ten different widths, are called *kote* or trowel, as their angled shape is reminiscent of a plasterer's trowel. Incidentally, tailors also use a similar tool, though with a longer handle, for ironing garments. The hot iron is passed over the glued joint, a wooden mould is inserted and its two halves prised apart with two wedges. Now the next strip is prepared, the wooden mould taken out, glue is applied to the inside of the outer ring and the second ring is pushed in from above. After the third ring has been glued in the core is smoothed with knife and sandpaper.

The container wall is polished with leaves that have been previously soaked in water.

Kaba-zaiku – vessels made of cherry bark

Ogasawara places the cherry bark that has been prepared with the scraper and cut to
size around the core, which has been fully coated with glue. He then presses the bark
on with the hot iron.

Cherry bark with different surface structures is used. If only the conspicuously protruding parts have been removed and the natural texture retained it is called *shirashi*, Cherry
bark that has been polished smooth is called *migaki-kawa*. Natural abrasives are used
for polishing, first with field horsetail (*tokusa*) and then with leaves from the Muku tree
soaked in water; Ogasawara harvests these in October from two trees planted in front
of his workshop.

Products – marketing

The Ogasawara workshop makes mainly the tea caddies called *cha-zutsu* and quivers
(*ya-zutsu*) for archers, usually in small series of five to ten. 200 tea caddies at a price
of 7000 to 100,000 yen and 80 to 90 quivers, which bring in 30,000 to 1,200,000
yen leave the workshop every year. About 60 % of production goes to specialist shops,
the rest to private individuals. Smaller quantities of large round boxes for tea-services
(*cha-bitsu*), small tea caddies without inner lids (*natsume*) and trays (*obon*) are also
made.

Father and son –
both very proud of
their precise quivers.

A special feature, and the workshop's best-kept secret, is the manufacture of containers for which the bark is not butt-jointed. The bark is not detached from the trunk, but rather twigs or trunks are cut to length and the wood is removed to leave the bark. The diameter of containers like these, constructed from the outside inwards, varies from item to item. The great attraction is the fact that there is no seam on the outside of the container.

The high precision of the containers is a particular source of pride. The cylindrical containers usually have a lid fitted on top. If this is quickly pushed on to a quiver or a pipe case and released again it will shoot back up into the air. This phenomenon, known as *nuki-komi*, in which compressed air makes the lid shoot off, is proof that the container has been precisely made.

Bill-hook and
half-finished spoon
on the wooden block.

Spoons

Ladles (*shakushi*) for soup and the flat spoons, somewhat reminiscent of spatulas, for serving rice (*shamoji*) are found in every Japanese kitchen.[15] From the 18th century at the latest there were villages in wooded regions where they were made as a principal occupation or sideline. Two mountain villages, in which almost the entire male popula-tion worked as spoon carvers until the post-war years, are Sotani and Shinohara in Yoshino, a remote mountain region south of the old capital, Nara. Sotani is the last of three hamlets in a narrow side valley whose steep slopes, some of them terraced, are still used by the inhabitants for cultivating vegetables. Before the war Sotani still had 60 households, today it is down to 17, and in ten years probably just a handful will remain. In this valley, unlike northern Japan, the spoons are not made of copper beech but of chestnut wood (*kuri*). Spoon manufacture plays a special part among Japan's wood trades for two reasons. The wood is worked on while it is still fresh, and not in a fixed workshop, but originally in the wood, immediately after it was felled. The spoon carvers usually worked in groups of four, who built a hut for themselves in the wood. This was used to live and work in, and had a fireplace in the middle. A good spoon carver, work-ing from five in the morning until ten at night, would produce up to a 100 spoons per day. They spent almost the whole year in the woods, returning to the village only three

160

times, at New Year, in May to cultivate the steep fields and finally for the Buddhist Feast of the Dead in August. The women brought fresh food every five days, and took the completed goods to the wholesaler's collection point on their way back. Once all the suitable timber had been used up, they moved on and built themselves a new hut. The great collapse came roughly in the mid fifties, when metal and plastic spoons started to take over. They were not just seen as modern, they cost only a tenth of the price.

Today only the 77-year-old Atarashi carves spoons in Sotani, in the fifth generation. He now works in a small two-storey workshop by the house, which like everything fights for its life with the slope. On the valley side, a spoon a good two metres long hangs as a sign to recognize his premises by. From his workroom, which is about the same size as the huts in the wood used to be, the view is over a few terraced fields and a great deal

The workplace –
Atarashi works sitting
at the chopping
block, his tools hang
within reach.

Sotani – a small,
scattered settlement
in a narrow valley.

The basic shape of the spoon is cut out with the bill-hook.

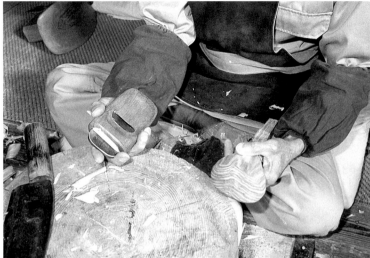

A short plane is used to smooth the underside of the bowl. The wood is moistened to make it easier to work.

The spoon is clamped between chopping block and stomach protector for work with the drawing knife.

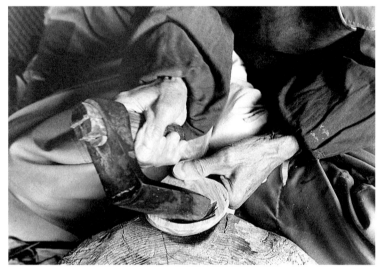

Atarashi cuts the bowl with a concave adze (naka-uchi).

of woodland a few kilometres down the valley. The wall behind him and to his left hangs full of tools. Many of these are very specialized, like curved drawing knives and adzes of various different radiuses. Atarashi cannot buy his tools, which have become very rare, from normal traders, but has them forged to his own specifications by a smith of the same age in the next little town. It is not certain how long he will be able to go on supplying them.

The diminishing supplies of wood are also cause for concern. Until 20 years ago his needs could be met on the spot, but now Atarashi has to order his chestnut wood from northern Japan. Kuri, whose hard and very weather-resistant wood used to be used for sills and floor timbers, is not very popular among foresters today, as the trees need five times as much space as Japanese cedars.

One foot is used as a
stop, the second
pushes the workpiece
against the floor so
that the bowl can be
hollowed out with a
special drawing knife.

The handle is
smoothed with a
straight drawing knife.

Atarashi uses only fresh Kuri wood. It would be much more difficult to work with dry. The chestnut trees, which are at least 60 years old, are cut to the required length and the sections are then split with a maul and cleaving iron (*wari-bôchô*) radially into two, four six or more parts according to diameter and end product. The basic shape is produced with two different bill-hooks, one for the coarse preliminary work (*kidori-nata*) and a second one for the finer subsequent stage (*ko-zukuri-nata*). Such bill-hooks with a sharpening bevel on one side only are also used for cutting off branches in the forestry trade. The spoon carver works sitting down, using a large 25 cm high chestnut chopping block as a rest. This is called *kozukuri-dai*, a "rest for fine work". The next stage is to round off the under-side of the round scooping part, the bowl of the spoon, with a very short, one-handed plane (*teppen-ganna*) that Atarashi makes himself. He holds the spoon at chest height with his left hand and works the plane with its shallow-set blade and concave sole in short, round movements. Then he works on the handle with a straight drawing knife (*sen*) that he pulls with both hands. For this step he clamps the spoon between the chopping block and a protector (*hara-ate*) made of chestnut or Japanese cedar that he has placed in front of his stomach. This stomach protector, polished by long years of use, is curved and thus fits the craftsman's body as he leans forward; it has a protruding section at the front where the end of the spoon handle is placed. Finally the hollow in the bowl is cut out with a short-handled hollow adze (*naka-uchi*). It has a round cutting edge, and it is worked in short thrusts. The hollow is then finished off with a drawing knife (*uchi-guri-ganna*) with a U-shaped curved cutting edge. The two handles of this tool, which Atarashi has in two dozen different radiuses, are bound with string so that the blade keeps its shape. Atarashi changes his working position again for this step, using his left foot as a stop and pressing the handle against the floor mats with his right foot. Finally the stomach protector is put in place again, the spoon clamped in and its handle, straight or slightly curved according to model, smoothed with the drawing knife. Atarashi piles the completed spoons carefully into little towers to air for a couple of weeks. So the product rather than the base material is dried.

Atarashi produces mainly three basic shapes in four sizes each. For the deep ladling spoons (*tsubo-shakushi*) and the "flat" spoons (*hira-shakushi*) the handle starts exact-

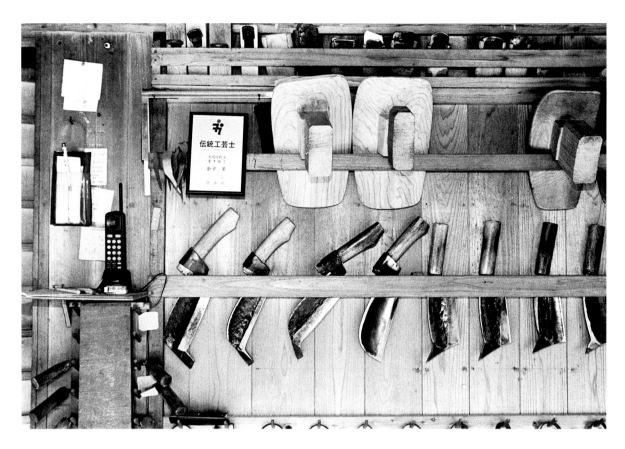

Tool wall with planes, stomach protectors and bill-hooks.

ly in the middle of the bowl, so they are symmetrical. The third basic shape, which was invented in Sotani a few generations back, is called *noshi-shakushi* (from the two characters *no* and *shi* in the Japanese phonetic alphabet; when written by hand from top to bottom the two signs look like the shape of this spoon); here the handle is placed obliquely to the bowl. As for some other traditional tools (e.g. kitchen knives or scissors), these spoons are also made in a left-handed version. Other items produced by Atarashi like fish-slices, scrapers and also beautiful large wooden bowls are more recent products that he makes using the same tools and similar techniques.

Examples of the major working steps.

Atarashi is now one of the last exponents of his profession. His products have rarity value, and so he commands a much higher price than his predecessors, at 1300 to 2800 yen per spoon.

The spoons have a particular charm: they show clear traces of the work that has been done on them, and seem primitive, but at the same time their slightly tapering, chamfered handles, angled bowl and curved outlines give them a rare elegance. Apart from cutting the timber to length, no measurements or marks are made, and yet they are high-precision products – as the piles show. Japanese craftsmen speak of *kan* or feeling – proportions and working have been completely internalized.

The completed spoons are carefully piled up to dry.

Moulds

The green tea that the Japanese prefer to drink is very bitter, and so from the earliest times something sweet has been handed with it to provide a balance of flavours. Many sweetmeats were made with wooden moulds (*ki-gata*), and in the Edo period countless motifs relating to a particular use emerged.[16] Moulds are made of cherry-wood, whose relative hardness and even structure mean that it can be finely carved. For the same reasons it is also used for woodcut blocks. No other Japanese city has such a rich sweet-making tradition as Kyoto; typical products are known as *kyô-gashi*. Today almost 2000 firms produce them, above all because traditional sweets are probably the most popular thing for tourists to take home as presents.[17]

The wide range of sweets and the moulds for them can be divided up according to size, motif and construction. There are small dried sweets, weighing only two to three grams (*hi-gashi*). These are made in long, narrow moulds and also handed with tea as an everyday occurrence. But the 40 to 45 gram, moist *omo-gashi* are usually reserved for the tea ceremony; these are made free-hand or with moulds.

The motifs can be divided into four categories. For the little pastilles they are usually flowers or leaves. They represent the seasons. The Japanese are particularly aware of their changing seasons, and this is a central theme in the tea cult. Particular care is taken that the sweets are appropriate to the season. In spring, for example, pastilles are served in the shape of cherry or peach blossoms. The second group uses Buddhist motifs (*butsu-ji*) like lotus blossoms or the wheel of learning (*rinbo*); such sweets, which are relatively large, are offered on the Buddhist house altars. There are special patterns, carp, for example, for particularly joyful events (*iwaigoto*) like weddings or the annual children's feast. Finally there are moulds carrying the emblems of shrines and temples that are used at their feasts.

Detail of a mould with two layers. The thin batten placed on top can be replaced several times over.

Receptacles and tools

166

The wooden moulds can be made with one, two or three layers. The paste is made of ground soya beans and sugar, and is then pressed into the mould with the diagonally cut underside of a conical palette knife made of hard Kashi wood. Moulds with two or three layers form a sculptural, three-dimensional body. In the case of three-layered moulds, the middle one is put in place, sugar paste is pressed into it and the excess is smoothed off. Then the middle layer is taken off again and the top one pressed into position. For this reason the middle layer has to be precisely the right size to allow the appropriate quantity of paste to be left when it is taken off for an appropriate shape to be created when putting on the third layer. Today there is scarcely a confectioner who is still master of this laborious technique, and demand is correspondingly small.

The Noguchi workshop

There is only one mould-maker left in Kyoto today, Noguchi, who was born in 1933. He works on the upper floor of a town house that is a good 100 years old. Located north of the main station in a side street running north-south, it is one of those buildings standing on a very narrow parcel of land that used to be typical of the city. On the right-hand side, a narrow corridor leads into the kitchen, the living rooms are all to the left of the corridor. In the room on the street side, two cupboards filled with about a hundred patterns, of which about half came from Noguchi's father, stand by the back wall. Noguchi has been carving moulds since his 18th year. He learned his trade from his father, who was born in Kyûshû and came to Kyoto via Hiroshima and Osaka, where he set up his workshop in the same house in 1919. Noguchi's greatest treasure are the three pattern books that came down from his father, who secretly copied the motifs, partly at the staging-posts on his progress through other workshops.

When Noguchi was apprenticed to his father after the Second World War there were still five mould-makers in Kyoto, and now he is the last. His son trained under him for a few years, but finally took another job because of a lack of commissions. As in many other trades, the profession can no longer feed its men.

Many traditional town houses have survived in the neighbourhood.

Manufacture and marketing

The cherry-wood is air-dried for at least three years, and then planed out and cut to size. The motifs are not traced on to the wood, instead Noguchi just draws the outline and the centre of the form if necessary, with a pencil. First he does some rough preliminary carving, checking the depth with a depth gauge made of sheet brass. Then the actual mould is cut. Noguchi holds the blank with his left hand against a piece of Keyaki planking, holding the short carver's knife in a fist grip. The thumb of his left hand serves as a support and guide for the knife; he has about a hundred knives laid out on his workplace in a semi-circle. He guides his knife away from himself at a relatively steep angle and turns the work-piece if needed. If it is a mould for dried pastilles, in which the many small identical motifs are very close together, he executes the same working step on all the motifs in succession, like a small series. If he were to complete the mould for one pastille fully and then turn to the next one, the shapes would turn out to be too different.

The sugar paste is pushed into the mould with the hardwood cone. The mould is worn down considerably from this. In order to avoid renewing the entire mould frequently, the preferred approach is to cut only the actual motif, placing a thin, pierced batten over the mould. Two little wooden dowels fix the batten, which can be renewed, if necessary, with relatively little effort.

Noguchi's father's pattern books.

Noguchi only works to order. Almost all his commissions come from a local specialist shop for confectioners' equipment. Although the shop of course builds in a margin, Noguchi does not try direct marketing, but likes to distance himself from the customer in this way. He is particularly annoyed by confectioners who come round on occasions with a competitor's pastille that is selling well and want to order an identical mould. He turns down commissions of this kind on principle.

Noguchi in his living-room. The glazed cupboards in the background contain some of his father's moulds.

Noguchi uses short chisels, which he fits into the handles himself.

169

Moulds

Combs

Combs have existed in Japan for at least 2500 years. The oldest known examples are made of bone and lacquered bamboo. But since the Middle Ages at the latest they largely have been made of wood. Boxwood is particularly popular (*tsuge*; written with the two characters for "yellow" and "willow"), a very hard wood with an even yellow colour that is very good to work and polish, slides through the hair well and has a pleasant feel to it. Three main groups can be distinguished: combs for caring for and cleaning the hair (*toki-gushi* and *suki-gushi*), for dressing the elaborate traditional hairstyles and finally the combs women put in their hair, which were often lavishly decorated with paint and carvings (*sashi-gushi*).

Jûsanya workshop in Tokyo

Tokyo was one of three centres for the manufacture of wooden combs, along with Osaka and Yabuhara in the Nagano Prefecture. The comb-makers' association still had 54 members here in the early 20th century.[18] The demand for high-quality wooden combs and thus the number of workshops declined rapidly as they were displaced by industrially manufactured celluloid, and later plastic, combs and by the disappearance of traditional hair-styles caused by the acceptance of Western fashions. Another decisive event in the history of this profession in Tokyo were the fires caused by air raids in the final phase of the Second World War, in which most of the workshops sited between Ueno and Kanda lost their valuable timber stocks. Today there are only two workshops left in Tokyo. One of them, and probably the last in which combs are still made by hand, is Jûsanya near the busy Ueno station.[19] Translated literally, Jûsanya means house number 13. This does not refer to the location of the business, which was established in

The area where father and son work is separated off by a wide Keyaki plank. Different models are on display in the showcases on the right, and the way through to the living quarters is in the background.

170

1736 and is now in its 14th generation. This is the story behind the name: combs are called *kushi* in Japanese. The two syllables of this term, *ku* and *shi*, have negative implications, as they evoke suffering (*kurushimu* = to suffer) and death (*shinu* = to die). But at the same time *ku* and *shi* are the Japanese words for the numbers nine and four. So the *kushi* sound sequence's negative connotations were avoided by adding the two numbers together and calling the business "13".

Jûsanya is one of the very few remaining premises combining shop and workshop, with manufacture and sales in the same rooms. The family lives on the top floor of the house, which was built after the 1923 earthquake. It is opposite the large Ueno pool, not 300 m from the station of the same name in the north of the city. Over the entrance, whose four sliding doors take up the full width of the building, the name of the shop hangs in the form of three gilded characters, with a large gilded comb above them as a sign. Only the area immediately behind the sliding doors is at ground level, the shop behind has a high floor covered with tatami mats. On the left along the wall is the narrow manufacturing area; the area in which Takeuchi Tsutomu, born in 1942, and his son sit to do their work, is separated off by a Keyaki plank with a particularly fine texture, standing on its end. When a customer comes into the shop they keep concentrating on their work, thus giving him time to get his bearings; but as soon as they are spoken to they are immediately at his disposal. The workshop's products are displayed modestly in a low wooden showcase at the entrance. A large wooden wall cupboard with a lot of small drawers, the black-painted sign from the previous shop, a little exhibition of a number of old comb forms and the obligatory shelf for the gods make it clear that traditions live on unbroken here.

The shop is in a busy street near the station. A large comb hangs over the entrance to mark it, and under it is the name of the workshop, Jûsanya.

Takeuchi only talks when the customers want him to.

Shelf for the Shinto gods.

About 200 blanks are held together with bamboo, ready for smoking.

Material supplies and drying

The Jûsanya workshop is one of the very few that uses solely indigenous boxwood. It is hard and yet flexible, feels pleasant against the skin and takes on a deep glow with use. It is obtained from a sawmill in the Kagoshima Prefecture on the southern main island Kyûshû that specializes in cutting boxwood. Boxwood is the preferred material for pieces in games and the seals that are used in Japan instead of signatures, as well as for combs. Box trees have been cultivated for their timber for generations, and are felled at the age of about 50. Thin discs are then cut from the trunks, which are a good 20 cm thick, and wedge-shaped blanks are cut from these. The blanks, called *kushi-ita* (comb-board) are bundled with bamboo hoops and smoked in an iron cupboard behind the workshop for about 15 days. Shavings and sawdust produced while making combs are used for the smoking. The process, similar to the one used in the manufacture of bowls, draws substances out of the wood, which means that it lasts better. The blanks, sooty at the edges, are then dried for five to ten years in the workshop. Long-term planning is thus essential in this craft.

Making a comb

After the blanks have been coarsely cut by hand and planed to the required thickness, the teeth are cut (*ha-biki*). For special products this is still done by hand, but a circular saw has been used for the commonest forms for over 50 years. Further work is carried out sitting at the little work-bench (*ishi-uma*, stone horse). It is a granite block with sides about 25 cm long, with one continuous groove or two short ones cut out of the top. The craftsman places several so-called heads (*kubi*) in these, usually made of hard Keyaki wood, and fixes them with wooden packing pieces and wedges. These primitive resources create a complex workplace that can be retooled in a few seconds. First a kind of file, a wooden bar with a wedge-shaped cross-section, tapering conically, and with ray fish-skin glued on both sides, is used to work the points into a conical shape (*ara-zuri*, coarse filing). This is done with a slightly rocking movement, even and rhythmical. The comb is placed at this stage in a rebate on the head, fixed with the help of a little wooden wedge. For wider models the wedge has to be loosened once and the comb moved. Then an iron file is used to work on the gaps between the teeth and a saw-like file to work at the base between the teeth (*gangiri* and *ne-zuri*). Fine work on the teeth

The smoked and dried blanks are prepared with the plane and coarsely cut.

(*ha-zuri*, filing teeth) is done with special files (*ha-zuri-bô*) made by gluing horsetail (*tokusa*) on to a wooden stick with a triangular cross-section. All the phases of work on the teeth are carried out from both sides, so the comb has to be turned several times. Horsetail, with its softly bedded silica molecules, was a widely used abrasive until the

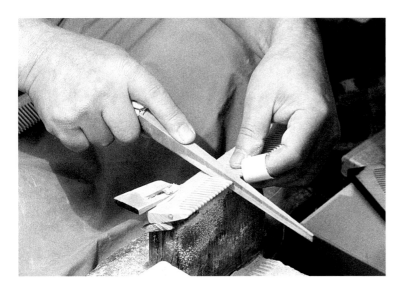

post-war years. In most cases it was replaced by sandpaper, but it is still used when working with cherry bark, as well as for combs.

Then the comb-maker applies a wooden template and draws the outline of the comb. He places the workpiece in the wedge-shaped groove of a vertical head set to the right of the work-bench, holds it with his left hand and cuts the arched top end out with a small coping saw. He repeatedly runs the narrow blade of his coping saw over an oil-soaked cloth he keeps on his bench in a thick piece of bamboo cane. The head on the bench is changed again and a "polishing rest" (*migaki-dai*) is inserted, a narrow rest that is covered with white felt at the top for a better grip and to protect the workpiece. The upper corners are bevelled with the coping saw before the edges are rounded off with a plane (*marumi wo dasu*). To do this the comb is pressed against the underlay with spread thumb, index and middle finger, and the right hand is used to draw a small block plane with a flat sole and without a cap iron. It might seem that it would be difficult to control such work on a curved surface with a flat-soled plane, but the craftsman manages to round off the comb astonishingly elegantly and excitingly, even on the small radiuses of the rounded corners. The final smoothing and polishing process is done with flat battens that also have horsetail glued on to them. Finally the teeth are brushed with

A wooden template is used to mark the outline of the comb.

a brush made of pampas grass roots and the comb is rubbed with camellia oil, which gives the boxwood a matt, silky sheen and a deeper colour.

The combs are made in small series of four or five. Thus each individual phase of the process is carried out on five combs in sequence. This is more efficient than making them one at a time, and at the same time retooling the bench relatively frequently and changing the tools avoids monotony and thus promotes concentration.

The master does not intend to shorten the elaborate manufacturing process with a total of over 70 steps. Creating the right shape for the teeth and smooth gaps without edges guarantees that the comb will slide through the hair without offering any resistance. The appreciation of a group of faithful customers who swear by his combs strengthens his self-confidence. His eldest son is following him into the profession.

The comb is put into a grooved batten that is wedged upright into the stone block. A small coping saw is used to cut the curved upper edge. The saw blade is repeatedly drawn over an oiled cloth stuck into a short piece of bamboo.

Shaping the upper edge with the coping saw. The comb is placed on the polishing rest for this purpose.

The edges are rounded off with a plane.

*The three most
important stages –
blank, after the teeth
are cut,
complete comb.*

Shapes and types

The Jûsanya workshop makes a wide range of combs in terms of use, form and teeth. They used to work almost exclusively for hairdressers (*kami-yui*) who dressed hair in the traditional Japanese way (*nihon-gami*). The combs for laying hair, made entirely by hand, usually have a long round handle set in the middle and tapering towards the end. These combs, now used only by Sumo wrestlers, the geisha apprentices called Maiko and for wigs, are now the exception. 98 % of the production is used for day-to-day hair care. There are several basic forms, strictly rectangular with teeth on one or both sides (*kataba* or *ryôba*), those with curved top and an asymmetrical outline. The combs most used by women (*bin-kaki-gushi*; broad combs with an asymmetrical shape running to a point on one side) are offered with teeth of three different widths, coarse (*araba*), medium (*aiha*) and fine (*ainaze*).

Combs are also used as utensils in certain Shinto rites. They are said to have the power to fend off negative influences. So the Jûsanya workshop has made combs for the shrines of Ise, which are renewed every 20 years. The combs for the accession to the throne (*daijosai*) of the current Tenno were also made here. The last used to be made of Isu wood, Takeuchi reports, the only indigenous wood that is almost black. Now box-wood is used for these as well. In terms of numbers such commissions, seen as an honour and distinction by the craftsmen, are of no consequence, but because they recur periodically they guarantee that shapes and techniques, some over a 1000 years old, are handed down faithfully.

175

Combs

Geta – wooden sandals

Old Japan had a wide range of footwear with an abundance of variants specific to pro-fessions and regions, usually open at the top and thus falling into the category of san-dals.[20] Various materials were used for this, straw and fabric, but above all wood. Of the variants made of wood it was above all the *geta* that were most widely available and are still worn in the present day. *Geta* are sandals in a plain shape consisting of a wood-en board with two bridges called "teeth" (*ha*) protruding on the under-side. On the top the foot is held by two straps meeting at a point at the front of the sandal. Usually the whole sandal including the bridges is cut out of a single piece of wood. The preferred material is Kiri (Paulownia), as it is light, cracks very little and feels pleasant in summer and winter because of its low heat conductivity. Other kinds of wood are used on rare occasions, Japanese cedar, for example.

Until the late fifties *geta*-making was a one of the most widespread crafts. There was a centre for industrial production in Matsunaga east of Hiroshima from the early 20th cen-tury, but little workshops were spread all over the country where sandals were made by hand.[21] Here the Shindô workshop is taken as an example of *geta*-making, as it is seen as the last workshop that uses no machines at all.

Smoothing the diago-nal known as hana *on the bottom; Shindô's feet are clasping the workpiece.*

176

The Shindô workshop

The Shindô workshop is in the little coastal town of Chichiwa in the Nagazaki Prefecture, directly on Route 57. Behind it is the cone of the volcano Unzen, which attracted attention in the early nineties when it erupted and caused several communities to be evacuated. Here Shindô runs a business that also sells footwear other than *geta*, some traditional, some tailored to local needs. The workshop is practically part of the shop area, though it is a good 50 cm higher and can be separated from the shop by glazed sliding doors.

Shindô is the second generation to run the workshop. He was apprenticed to a *geta*-maker in the city of Nagazaki for four years from the age of 15, and also lived in his home (*sumikomi*). His father enjoined him to spend this time in full, and not even to come home if he should die. Shindô says that his master was a well-known *geta*-maker, strict, but an outstanding craftsman. Shindô came back to his father in 1955, and worked with him for ten years. Until the early sixties there were still two sandal-makers other than Shindô in the little town, as it was such a common profession. The decline began in the mid fifties with the rise of cheap shoes in fabric and plastic. But the demand never dried up completely, and Shindô never stopped making sandals. It was about 1975 that Shindô started to sell more to customers from elsewhere – his craft had become a rarity within a mere 20 years. He acquired a national reputation from 1985 as a result of reports on radio and television. In 1990 the Shindôs built their new shop on the truck road, a traditional two-storied wooden house with tile roofing. The volcano and the countryside around it attract a lot of visitors, and the road is very busy at weekends in particular. The sale of *geta* almost doubled when the new shop opened. The other shoes, which take up almost half the display area, represent only five per cent of sales. Shindô is sure that he only sells so many *geta* because the customers can see him making them.

Craft shop – slippers for the locals round off the range.

Workshop on the
trunk road, with the
Unzen volcano in the
background.

The small workshop on the narrow side of the shop has an elevated board floor. Shindô sits on a cushion on the floor; in front of him about a metre has been cut out of the floorboards. The shavings drop into this opening, and this is where the work-block called *kiban* is placed, a short cedar pillar with a square cross-section. It protrudes a few centimetres above the boarded floor, and an iron bracket is fastened to the top. According to the work-phase, this bracket serves as a stop against which Shindô pushes the workpiece with his feet, or different wooden rests are placed in the slit between the bracket and the top of the pillar and held in place with a small, moistened chunk of wood that slides in at the back. These rests, called *dai*, also have a stop (*tome*) on their top side, against which the craftsman presses his workpiece.

On the wall behind the craftsman are the rows of planes in rank and file, and above that are racks containing many chisels; their blades are usually in a protective wooden sheath. To his right is a tub of water and the whetstones, above the window is the obligatory shelf for the gods, and next to it hang the templates made of thin sheets of wood that are needed for sandals in various shapes and sizes.

Material

Until about 1970 Shindô was able to procure his Kiri wood locally, he bought most of it from peasants and also felled the trees himself. He cut the trunks to length by hand and split the off-cuts into two or four logs according to the thickness of the trunk (*han-bun-wari* or *yotsu-wari*). Then the basic shape was roughly cut with a hatchet. He then stacked these blanks called *makura* into a skilful pile a good two metres high in the open air. The rain washes certain substances out of the wood (*aku wo toru*), which makes it more durable; it also dries faster when the pile is dismantled after two months and the blanks are piled up again in a covered, shady place. They are stored here for at least four months. 30 years ago Shindô gradually went over to buying his blanks from the timber trade. He gets them from a dealer in Nara who specializes in Kiri. A pair of blanks for the smaller women's sandals costs between 1000 and 5000 yen, and for large men's blanks in close-grained wood with vertical annual rings he has to pay up to 10,000 yen.

Working phases

The example of sandal-making described here takes a women's sandal model with a round heel, called *ato-maru-dai* in Japanese ("round at the back").

The dried wooden blocks are first prepared for further work with bill-hook and roughing plane (*kidori* and *kanna-gake*). Shindô then places a thin wooden template of the kind he has for all normal sizes on top and marks the length with two pencil lines (*tengata-tsuke*). The block is now cut to length slightly on the diagonal at the heel (*kakato-kiri*). When cutting, his left foot rests half on the block and he pushes it against the stop with his heel.

The top of the sandals is not completely flat but slightly concave, by about 3 mm over the full length. To create this hollow (*sori-tsuke*), the wooden block is pushed against the stop with both feet and material is worked off across the grain with a wide chisel. This chisel is used only by *geta*-makers and has some striking qualities: the cutting edge is kept slightly oblique for a "pulling cut", and the chisel has nail-shaped extensions on both edges. They make contact with the workpiece, function as a stop and thus prevent the blade from slipping into the soft material. The concave surface is then worked with a plane with a correspondingly curved sole.

The next step is to cut the "teeth" on the bottom of the sandals. A wooden template that looks like a little wooden frame is used to mark the position of the teeth (*hagata-ate*) and then their depth is marked with a marking gauge. Shindô uses a large hand-saw (*abiki-noko*) to make four incisions. He introduces his hatchet into these incisions; by exerting slight leverage with the hatchet (*aikagi*) he splits the wood at the bottom of the incision, which makes it easy to remove the material between the teeth. The diagonal at the front on the under-side of the sandals is also created with the hatchet (*saki-otoshi*). Shindô now places the template on the top and draws the outlines of the sandal with a pencil.

The hollow on top of the blank is worked with a special chisel whose two edges have needle-shaped extensions.

Marking the teeth with a template.

Now the two teeth on the sandals are elaborately smoothed. Another wooden rest is placed on the work-block and fixed at the side with an iron clamp. Shindô puts his blank into this diagonal support; his left foot rests on the workpiece while he bends forward and uses the full weight of his body to take a shaving off the top of the sandal, using a levering movement of the two-handed knife, which is a good 30 cm long (*hama-kiri*). Then the corners are cut out with a long narrow gouge (*hama-no-nagashisuki*); this tool has a needle-shaped extension on its left-hand side to guide it. Finally the ground is worked with a broad chisel (*jûnô*). Shindô always sharpens the tools again before this final process and rubs them over a ball of cotton wool soaked in peony oil. Sharp tools are a must with the soft Kiri wood that is being used here in particular: "If it does not

179

The diagonal on the under-side is coarsely worked with the hatchet.

The wood at the top of the teeth is cleaned with a two-handed knife.

Holes for the straps are drilled with a wooden brace and spoon bit.

Planing the slightly curved sides.

shine, the tool will not cut" (*hikari ga nai to dôgu ga kirenai*). Shindô takes of wafer-thin shavings with a slight hiss, thus creating a mat silky surface.

Shindô cuts the two rear holes with a home-made brace with a spoon bit (*yoko-ana*). Then he planes the front diagonal on the bottom (*hana-kake*). The curve on the side of the sandals is made with hatchet and a plane with a concave sole (*yoko no maru kezuri*). Shindô rounds heel and toe with the two-handed knife (*marume*). To achieve the necessary pressure he presses his lower thigh against the back of the knife. He has previously wrapped a white towel round his lower leg.

Shindô puts the two sandals one on top of the other and smiles. Even though the outline was drawn only once and no further measurements have been made, the differences in size are less then half a millimetre.

Shindô is proud of his accuracy. When he has finished the sandals they vary by less than a millimetre.

Smoothing the rounded heel. Cloth is wrapped round the lower leg that Shindô is pressing against the knife.

Finally the front hole is drilled (*mae-no-anahogashi*) and a hollow cut out underneath with a chisel to make room for the knot in the strap. The top is smoothed with the plane, the edges are slightly bevelled and then Ibota powder is applied and rubbed in with a cylindrical brush made of willow roots. This surface treatment gives the sandals a slight glow (*tsuya-dashi shiage*), and also prevents them from getting dirty.

The two straps are knotted at the bottom; the knot at the front is then also protected by a small metal cap fixed with two little nails. A leather strip is attached with a few nails to protect the front edge of these soft-wood sandals against rubbing and damage.

It is fascinating to see how much effort goes into smoothing the wooden sandals. A perfect surface is not just aimed for on the visible sides and on top, it extends to the area between the teeth, which cannot be seen at all when the sandals are in use. The sequence of work, internalized and entirely in Shindô's blood and bones, leads to astonishingly precise products.

It is the craftsman's nimble feet that make it possible to work without any fixing device. A whole series of different foot positions come into play; the foot holds the workpiece by pressure from above, pushes it against the stop, and sometimes the toes are even used like claws to take hold of the workpiece.

Basic forms and uses

Here only the commonest shapes that Shindô now makes will be described. The simplest design is called *masa-geta*, "straight geta". The sandals are rectangular in format and have two narrow bridges on the bottom. In order to make best use of the timber, a special type of saw was invented in the 19th century enabling two sandals fitting inside each other to be cut from a single block. First the wooden block was cut to the level of the teeth on the two top ends. A *sashi-noko* (literally insertion saw) was used to create the 90 degree turn. Seven saw teeth are set at right angles in the middle of the narrow blade, which is about 40 cm long. This saw exists in three versions, with teeth of increasing height. Once the cut was 5 mm deep after inserting the three saws, work could continue with the usual pad saw (*itoko waki*). To make sure that the saw could be guided precisely the craftsmen used to place a small mirror behind the wooden block. This meant they were sure that the saw did not run off at the back either.

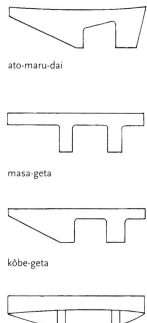

ato-maru-dai

masa-geta

kôbe-geta

taka-geta

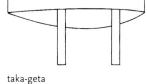

Drawing of various geta types.

Shindô scarcely ever uses this process now, which is elaborate even though it saves material.

The shape known as *kôbe-dai* is probably so called because it emerged in the port of Kôbe. The rear tooth is set back, the front one runs diagonally to the toe. This variant also exists is a version for men and women.

The little *tanjô-geta*, literally "birthday *geta*", are often given to children on their first birthday.

For the variant called *taka-geta* or "high *geta*" two grooves are cut into the bottom of the blank and thin strips of hard wood are inserted into them. Cooks like to wear *geta*-like this, especially in sushi restaurants with their wet floors. One advantage is that the bridges can simply be renewed.

Shindô makes about 1500 pairs of *geta* per year. He works every day, unless he is stopped by a wedding or a funeral. Formerly, when he was serving just a small local market, people used to buy their new *geta* before New Year or the Feast of the Dead in early August. Today a lot are sold to tourists or bought by local people as a gift when visiting friends. People still like to wear them on special occasions. They are part of the outfit for many traditional professions like wrestlers and geishas, but they are as popular as ever with cooks and crooks.

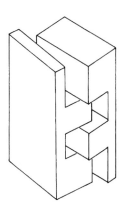

Economic use of
material for
masa-geta.

A master's feet.

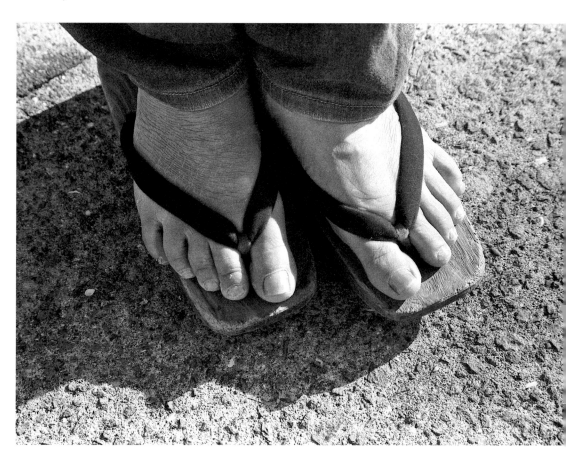

Making tools

There are four large centres for making cutting tools in Japan, Sakai near Osaka, Seki in the Gifu Prefecture, Miki not far from Kobe and Sanjô south of Niigata. While Sakai and Seki are particularly known for their knives, most hand tools for the woodworking trades are made in Miki and Sanjô. The tradition started 350 years ago in Sanjô and nearby Yoita with nail-smiths, then came the saw-smiths, and in the second half of the 19th century individual smiths started to specialize in making chisels and plane blades. The area came to be the leading centre for tool manufacture supplying a large number of trades, and this meant that from the late 19th century a lot of firms set up alongside the smiths making tools or parts of tools from wood.[22]

Two of the many workshops here – there are still over 70 – have been chosen because they are of particular importance for the woodworking trade, in fact by making planes and snap lines. Both are used in some form or another in almost all the timber trades. In comparison with other trades presented here the plane and snap line makers are relatively young trades with their 100-year-old tradition. But fundamentally tools, alongside buildings and containers, have been one of the main uses for wood as a raw material from the earliest times.

The planes in Inomoto's workshop are piled in pairs on top of each other on little trolleys.

Planes

Unlike Europe, where they have existed since Roman times, planes did not appear in Japan until the 15th century, and were in general use by about 1600.[23] Until then surfaces had been smoothed with a spear-shaped tool (*yari-ganna*). This has a slightly curved cutting edge at the end of a long handle; it is held in both hands and drawn over the wood to take off a shaving: this process leaves a smooth but slightly ribbed surface.

184

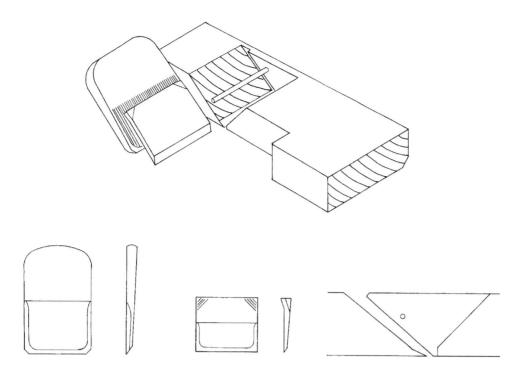

The actual plane with a fixed blade probably went back to continental models; it allowed surfaces to be entirely plane for the first time. The body of the Japanese plane, which is drawn towards the craftsman's body, is much lower than that of its European relative. The plane is made from a single piece of wood, and like many traditional Japanese products it is appealing because of its highly reduced cubic shape; it does not have a "horn" to get hold of, nor the hand protector behind the blade fitted to some continental European models. The very thick blade, which is slightly conical and forged in two layers, is set in grooves cut into both sides of the throat. Since the second half of the 19th century – following Western models – smoothing planes in particular have been fitted with a chip-breaker (*nimai-ganna*, literally plane with two blades), until then planes had a single blade and no chip-breaker (*ichimai-ganna*).

Originally carpenters and other timber craftsmen bought only blades; they fitted an appropriate handle to chisels, and they fitted the blades into the plane bodies themselves. Then Tazawa Torakichi is said to have set up the first workshop, in Sanjô, that specialized in fitting plane blades. This was the beginning of a new profession, called *daiya*. *Dai* is the term for wooden plane bodies, and *ya* the character for house or profession. Because of the building boom associated with Japan's rapid economic development and the great demand for hand tools the new profession must have spread rapidly. 200 such workshops are recorded for Tokyo alone in 1935. Specialization went even further, soon there were workshops for particular types of planes like for example grooving planes, rebate planes, moulding planes or the tiny ones called *mame-ganna* (literally: bean plane). The blade was fitted in only roughly (*gobu-shikomi*) until well into the thirties, after the Second World War they went over to making planes that were ready for use (*sugu tsukai*). The industrialization of the building trade and the hand machines that emerged from 1960 reduced the demand for hand planes to less than a tenth of pre-war levels. And so *daiya* are now only to be found in the two great tool metropolises.

Sketch of a plane; below: cap iron, plane blade and section through plane throat.

Material

Kashi is used almost exclusively for plane bodies, a kind of beech that provides hard, tough wood with an even structure. It can be worked very precisely and unlike oak, which has a high tannic acid content, it does not react with the iron. Usually it is *shiro-gashi* or white Kashi (relative weight density 0.83), rarely the somewhat softer and lighter red Kashi (*aka-gashi*). Kashi grows in the Kantô region above all. The 40 to 70 year old trees are felled between December and early March; their trunks are 30 to 50 cm in diameter. Wood from older trees is not as sought-after, incidentally, as its heart takes on a grey-

The plane blade is positioned and its width marked.

Cutting the groove on the throat of the plane body; this is where the blade will be inserted later.

Marking the gradient of the blade.

The mouth is opened, the left foot pushes the workpiece against the work surface.

ish-black colour. The sawmen call it *botan ni naru* ("becoming peonies"), as discolouring of this kind looks like peony flowers on the end-grain. There are sawmills that specialize completely in cutting Kashi. One of these is immediately behind the station in Sanjô, the Tomosaka sawmill. Here about 350 m³ per year are cut into planks with a band-saw, and then further trimmed on the circular saw as blanks for plane bodies, hammer handles and hafts. Until about 1920 wood for plane bodies was still split (*wari-dai*), which produces material that lasts better and does not warp. But labour intensity and low exploitation levels brought about a shift to cutting the wood for the planes (*hiki-dai*). Today the plane workshops buy cut blanks from the sawmill; these are called *ara-dai* or "rough plane body" (303 x 95 x 42 mm).

The Inomoto workshop

There are still about 25 plane workshops in Sanjô, mostly one-man businesses special-izing in a particular kind of plane. The Inomoto workshop has shifted its product range in the last 30 years from small or specialized planes to the "flat" planes (*hira-ganna*) that are most used; flat because they have a flat sole and are for work on straight areas. Most other kinds of plane like grooving, rebate and moulding planes have been forced out by the rise of machines, especially spindle moulders and overhead routers, and of mechanical pre-fabrication in the building industry, but there is still a relatively steady

The ground is cut for the blade bed.

The throat grooves are finished with the saw.

The blade is drawn over an oil-soaked wad and placed in position.

demand for smoothing planes. Inomoto, born in 1938, has a small workshop on a side street in the old town centre. On the top storey of the wooden building, which is a good ten years old and clad in sheet metal, the blanks, piled on edge and cross-wise, dry for two to three years. On the ground floor the machine room runs into the area where the blades are fitted into the plane bodies by hand. Half-finished and finished planes are stacked up in front of the work-bench. Since about 1960 the preliminary work has all been done mechanically, so the blank is first planed, cut to length and the throat is then roughly cut out with a router. The blade is built in at an angle of about 39 degrees. In Japanese the angle is given as a gradient ratio (*hachibu kôbai* – a gradient of 8 *bu* or 24.2 mm over a length of 10 *bu* or 30.3 mm).

Excess material is removed with a Shinogi-nomi, then the plane body diagonal is slightly roughened with a file.

187

Inomoto works cross-legged at a 60 x 50 cm hardwood working surface that has been let into the floor. Two little wooden blocks have been let into it; they protrude by a few centimetres and are called "claws". Inomoto pushes the blank against one of these stops according to the work phase. His tools, a large number of chisels, a few files, planes and drawing instruments, are in racks or hang on the wall behind his back.

The bottom of the diagonal blade bed is smoothed with the offset kote-nomi.

The hole for the cap iron is drilled.

Fitting the iron

Knocking in the steel pin for the cap iron.

Inomoto places the iron on the body and marks its width with a pencil. Then he fits the blank into a router to cut (*shikomiki*) the throat with its slightly curved iron bed and the two grooves in the plane block (*osae-mizo*). He returns to his workplace and marks the position of the blade with square and template (*kôbai wo dasu, sumitsuke*). He now places the blank on a rubber mat and pushes it against the side stop, puts his left foot on it and works on the sole mouth with a wide chisel. Next the blank is turned over and the block is worked on. In this workshop most smoothing planes are given a mouth called *tsutsumi-guchi*; the bed, in other words the diagonal behind the blade, does not run straight through to the sole but is matched to the diagonal of the plane blade. This means that the blade's bevel is supported at the rebate, which is intended to guarantee vibration-free work. Planes with a straight iron-bed are called *futsû-guchi* ("normal mouth"). Sooner or later the *tsutsumi* mouth, which is considerably more laborious to construct, will become a normal mouth, when the sole of the plane is repeatedly trued up and worked off.

Setting the blade and cap iron.

188

The two grooves for the conical plane blade are finished with a saw and a narrow mortise chisel before the actual fitting takes place. To do this the plane-maker draws the back of the plane blade over an oil-soaked wad he has next to him in a piece of bamboo. He then places it in the grooves on the throat of the block, then immediately takes it out again. The oil marked all the points on the iron bed at which the blade touches. He now knows where he has to chisel off more superfluous material to make sure the blade lies as flush as possible and thus makes it possible to work without vibration. He repeats this step up to 30 times to ensure that the blade fits perfectly.

Now the cap iron is fitted (*uragane awase*). If necessary some material is chiselled off the throat's sides, the hole for the chip-breaker pin bored and the pin knocked in. The two top corners (*mimi*, "ears") of the cap iron are slightly bent, Inomoto adjusts them with a few hammer blows at a small anvil and then fits blade and cap iron. Something that would have been unthinkable even 20 years ago is the rule today: the plane-maker trues up the plane. To reduce friction resistance he works with the short scraper plane and a chisel to make the sole slightly concave and even sharpens the blade.

Planes with interchangeable blades (*kae-ba*) now account for almost 80 % of production and 60 % of sales. As these blades are made mechanically and thus practically without any differences in dimensions no individual fitting is needed any more, the work seems to proceed at a great pace. So far they are nowhere to be seen at temple and teahouse sites, but they are now almost inevitable at the many prefabricated building sites. Inomoto earns 1900 to 7000 yen for fitting a blade to a traditional plane, deducting material costs of just under 500 yen. He receives only 1000 yen for planes with interchangeable blades. Inomoto fits about 10,000 plane blades per year, ten years ago when his father and younger brother were working with him the workshop still had an output of 25,000 planes. With isolated exceptions commissions come from about 15 wholesalers in the region.

Plane ready for use.

Snap lines

Carpenters use a snap line – called *sumi-tsubo* or "ink-pot" in Japanese – to mark their wood out. It would be scarcely possible to draw particularly long straight lines or mark off the unsquared beams often used in the roof frame without it. The importance of this tool for the Japanese building profession in particular is shown by the fact that along with the carpenter's square it is placed on the altar at the numerous building rites (ground-breaking ceremony, start of working process, setting up the first column,

placing the ridge purlin). There are special ritual tools for this purpose, elaborately decorated with lacquer and fittings.

Unlike the West, where marking out used to be done with two separate tools, a reel with a thread wound on to it and an ochre box, they are part of one and the same tool in the Far East.[24] A little wooden wheel is set into the rear section held in the left hand. A silk thread is wrapped round this little wheel (*tsubo-kuruma*, "pot wheel"). The head of the tool, which is considerably broader and usually roundish has a hollow known as a "pond" (*ike*). In this is a wad of cotton soaked in Indian ink through which the thread is drawn. The end of the thread is fastened to a little wooden peg with a pin (*karuko*). This pin is pushed into the workpiece; the string is then unwound as needed, which pulls it through the ink-soaked cotton wad, and stretched across the workpiece. To make the actual mark, the string is pulled upwards with the index finger and thumb of the right hand. When it snaps back the thread leaves a fine line of ink on the wood. Japan has had snap lines of this kind since the 8th century, they probably came from the Chinese mainland with the architecture. Hard Keyaki wood is used for this tool as it is resistant and easy to carve. Snap lines have differed slightly in design throughout the ages. There are also some local variants for certain professions, like plasterers, for example. A form emerged in the 18th century in which the front section with the hollow is decorated with a semi-plastic carved stork and a tortoise. These two creatures are thought to bring luck because they live for so long, and they are still the standard design for high-quality snap lines. If they are to function flawlessly there must be scarcely any play in the little wheel, otherwise the string gets tangled when being wound back; the two holes the string is drawn through must also be perfectly aligned.

Carpenters used to carve their own snap lines, but in the late 19th century a profession emerged that did nothing but carve snap lines. The highest concentration of these was in the tool metropolis of Sanjô. The building boom after the lost World War brought a last heyday for the profession; its numbers have been slowly dwindling since 1960. Today just four workshops remain, but only the Tamaki workshop has an heir. The blanks are now largely machine made (with a copying lathe and router), though the carving is still done by hand. The 77-year-old Tamaki, who started to work independently at the age of 18 after a five year apprenticeship, shows the perfection and speed this specialization can bring. He has about 50 carving tools laid out radially around his workplace, and he picks the right one up almost without looking. When carving he pushes the snap line with his feet against a stop on his round work surface, which is made from a section of a tree trunk.

When the carving is completed, the hollow for the cotton wad is painted black before the tool is soaked in an oil bath. Products from Tamaki's workshop almost all go to about

The father's workplace.

50 wholesalers in the region. To compensate for diminishing demand, the workshop has also produced blanks for false arms and legs for the past ten years. The plastic snap lines that emerged around 1960 do not cost even a tenth of the traditional version carved in wood, and because of extreme rationalization in timber construction, down to completely automatic prefabrication, carpenters rarely have to mark off by hand. Even so, almost all carpenters and joiners have this tool in their boxes, though only very few of them use it every day.

The workpiece is held with both feet and pushed against a stop.

Tamaki, father and son.

玩具

Toys and musical instruments

Nô masks
Wooden toys from the Yonezawa region
Boards for Go and Shôgi
Drums
Shamisen
Koto – long zithers

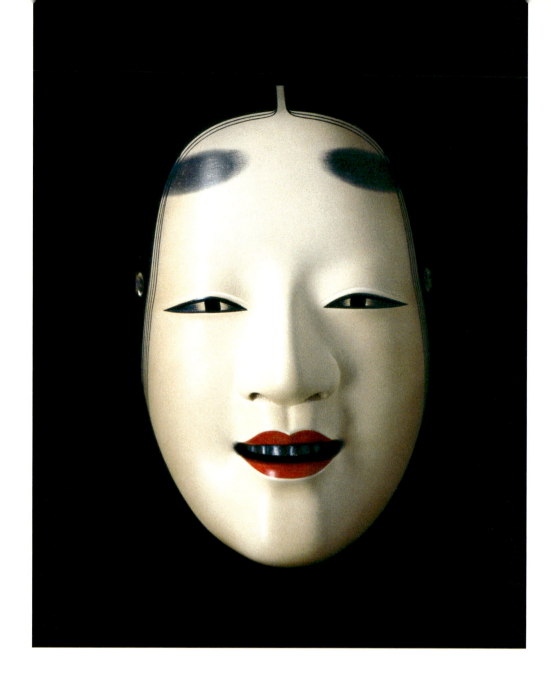

Ko-omote, *mask of
a young woman.*

Nô masks

Nô – Japan's classical theatre form

Masks have a long tradition in Japan. They were first worn for ritual dances at temples and shrines. Masks of this kind dating from the 8th century are also stored in the famous Shôsôin treasure-house in Nara. Masks then became the most important stage properties of Nô, the traditional classical Japanese theatre, established by the 15th century and passed down almost unchanged since then. Here both performers wear masks; the actors first contemplate their mask, which embodies a particular character, in a so-called mirror room (*kagami-no-ma*) set up especially for the purpose. In this way they assimilate their role before actually putting the mask on.[1]

Since the Edo period at the latest there has been a particular trade specializing in Nô masks; they are called *nô-men-shi*. A good 50 basic types are divided into eight categories, like for example the god of long life (*okina*), supernatural beings (*jinki*), man (*otoko*) and old man (*jô*).

Muneharu Nagazawa lives in Fukakusa in south Kyoto. The streets in this quarter, which are narrow anyway, are edged with flowerpots, motor-scooters and also the odd washing machine. Nagazawa's workshop, or should we say studio, is on the top floor of a house that is a good 40 years old. In his eight-mat room, Nagazawa is surrounded by faces from this and other worlds: a number of old masks are hanging on a lintel, under these several animal masks, large figures of Buddha stand in two corners of the room, and there are many half-finished masks on shelves. On the floor is a board just under one metre long and about 50 cm wide. The carver sits on this board and presses the mask with one or both of his feet against a short batten attached to the end of the board. Beside the workplace are several low wooden boxes containing serried ranks of carving tools.

Nagazawa is the second generation of his family to work as a mask carver. His famous father died in 2003 at the age of 92. He was so to speak placed under government protection in 1979, as the "holder of a traditional technique indispensable for the conservation of cultural properties". The middle son did not choose mask-making as his profession immediately, first working for twelve years as a graphic artist. He started to carve when the whole graphics industry went into a spin during the first oil crisis, followed by a paper shortage. His two brothers also work as mask carvers.

Nagazawa in his workshop, flanked by two old Buddhist sculptures.

The outlines of the face are repeatedly marked in pencil.

Contours are regularly checked with cardboard templates.

Making the masks

Cypress wood (*hinoki*) has been used for masks in Japan since the earliest times; it is light, resilient, easy to work and is avoided by pests. Nagazawa prefers old wood, although it is harder and thus more difficult to work. He buys his raw materials from demolition firms, in the form of large posts from old buildings. He has now laid in such copious supplies that he will scarcely be able to use them all.[2]

The basic shape – called *aratori* – is made with saw, axe and a broad chisel. Even before embarking on the first steps the carver repeatedly draws a middle line, which he establishes with the help of a compass, and marks the lines of the face. He works with chisel and a mallet for as long as he is taking off coarse shavings, pushing the blank against the stop on his workplace with his feet.

Nagazawa buys his carving tools from a smith near the National Museum, and fits them into a cypress handle himself. As work proceeds, they soon get finer. He sits with his legs crossed, holds the mask with his left hand and takes increasingly thin shavings off with the tool in his right hand. The carver has a whole set of cardboard templates (*katagami*), so that he can follow the old mask he is using as a model precisely, and

Carving the eyes.

to check the dimensions. They are largely outlines of all the important points on the face. The back of the mask (*ura-men*) is not hollowed out until the basic features have been carved at the front. Finally the mouth is opened up. To do this, a very fine padsaw, called *ito-noko* or thread-saw in Japanese, is introduced through a drilled hole. The mouth and eyes are worked on with a *ken*, a pointed scalpel with a sharpening bevel on one side.

The back of the masks is then coloured with black Urushi lacquer. The front is primed five times with shell limestone dissolved in animal glue. After intermediate smoothing, templates are again used to mark reference points like the hair-line and the eyebrows, before the mask is painted.

*Making indentations
on the back of the
mask.*

Nô masks

Status of the craft

The profession enjoys a status somewhere between craft and art. The products are highly expressive and yet the craft is essentially reproductive; there are about 250 mask shapes, Nagazawa mentions 70-80 basic shapes and variants, and they were all established by the 17th century. They are called *ko-men*, literally "old masks". The carver works from these, and the cardboard templates are made from them. Thus creativity is kept within the strictest bounds. The back of the mask does allow some scope, here each carver has his own handwriting. Nagazawa makes additional inner and creative scope for himself by occasionally carving quite different masks; he shows a tortoise's head and a grinning monkey.

One particular attraction of mask carving is its diversity: the *nômenshi* is first a sculptor, but he also has to know how to work with metal (*chôgin*) (gilded metal overlays are often fitted to eyes and teeth) and he has to paint his work (*saishiki*). For Nagazawa the fascination lies above all in the fact that he is able to take a block of wood and produce a form that seems to have a life of its own.

A searching gaze – the carver seems to be communing with the emerging face.

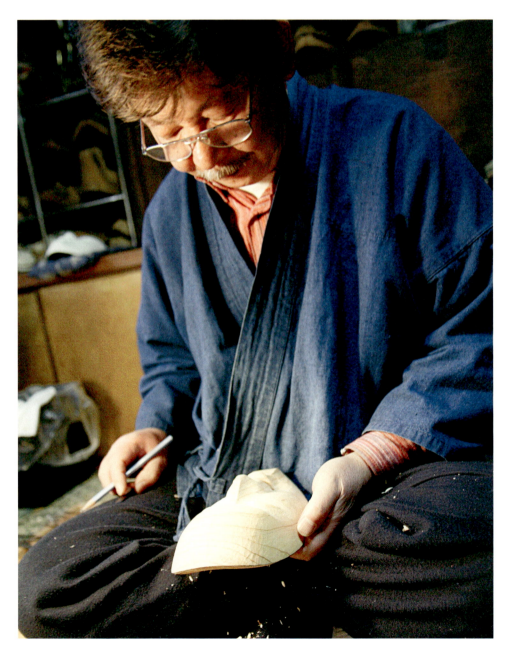

Today only a very few masks from Nagazawa's workshop are actually used in the Nô theatre. Most of the commissions come from collectors, who pay an average of 600,000 to 800,000 yen for a mask. Masks by the father, who has not been carving for a few years now, command a much higher price, about four million yen.

*The front and back of
a finished mask.*

Wooden toys from the Yonezawa region

Traditional wooden toys are made in many Japanese regions, often in resorts or places of pilgrimage like Ise where toy-makers settled at an early stage because there was a constant stream of visitors.[3] The region around the northern Japanese city of Yonezawa has a high concentration of toy-makers, and also a wide variety of products. Two workshops and the techniques used there will be discussed here.

The pole is smoothed with a knife before carving.

Sasanobori – Birds with plumage made of shavings

The Sasano area south of Yonezawa is known for its carved birds and other work using shavings. There is a term *ittô-bori*, "carving with one knife", and in fact a single knife is used for almost all the work. The local people identify this craft tradition as having started very early: even the original Ainu population is said to have practised it more than 1000 years ago. In the 18th century, carving was promoted as a winter sideline in this snowy region by the tenth feudal lord of the province. The oldest surviving work dates back to this period. Originally these products were sold mainly at the festival of the nearby Kannon temple on 17 January. There were birds, and also carved flowers like chrysanthemums, camellias and peonies, which were used to decorate Shinto and Buddhist altars in winter. The work went on sale in a Tokyo shop for the first time in 1921. Since then these carvings can be found almost all over the country, no longer a limited local product but a hallmark of the region as a whole.

Toys and musical instruments

Toda workshop

Today 28 carvers, mainly farmers, have formed an association to carve toys as a winter sideline. Only the Toda family, who can point proudly to a tradition going back six generations, gave up farming in 1969, rented their fields out and made carving their main source of income. They run a workshop and retail outlet near the Kannon temple. About a quarter of the premises is the actual working area, with a higher, boarded floor; here Toda carves, a square fireplace with a pot-hook in front of him.

All the products are made of young local timber cut when clearing the forest. It is about 2 to 15 cm in diameter. Toda uses only two kinds of timber, about 80 % soft Koshi-abura, which is ivory-white throughout, and Enju, which is considerably harder, with light brown heartwood. In October and November they go into the wood for a week and fell the trees themselves. The little trunks are cut down to just under two metres and then debarked with a drawing-knife within a few days, before being dried upright outside for a month. After that they are stored upright in a shed for another three to six months.

Working steps

The coarsely debarked poles are first trimmed by taking off thin shavings all round with an obliquely honed knife (*chijire*). To do this, Toda sits on the ground and holds a long pole firmly under his left arm. The actual cutting knife (*saru-kiri*) has an unusual shape reminiscent of a bill-hook. The broad blade has two sharpened bevels, one along its long lower edge and a second oblique one at the end of the blade, for finer trimming. The bird's head is made first, with a few brief cuts; the pole is turned at the same time. The deep incisions for beak and talons, which require greater strength, are cut with the long cutting edge. Then the elaborate feathers are created with the short oblique cutting edge making fine, curved shavings. Great skill is needed here to avoid detaching each shaving, rather than leaving it in place. It takes about 50 cuts to create a wadded mass of feathers. When the bird is finished, Toda measures the length with square and thumb, then cuts the bird off the long pole, so that he can start on the next one immediately.

Toda at his workplace.

The carving starts – the timber is bevelled off at the head.

The family division of labour means that Toda's wife draws the talons, beak and eyes with water-colours and Indian ink. Finally she adds her husband's name to the work, impresses his seal on the bird and packs it in tissue paper in a cardboard box with a band round it.

The thumb on the back of the blade means that pressure can be carefully controlled.

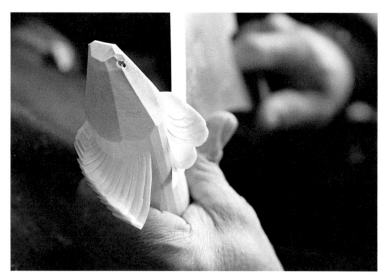

The feathers are made by a series of incisions using the oblique edge at the top of the blade.

The most important product of the Toda workshop are hawks, which have the longest tradition among the many shapes. They are made in sizes from just about 3 cm to over a metre, and are called *otakapoppo* – *taka* means hawk and *poppo* is the Ainu word for toy. These hawks occupy the middle ground between popular belief and toys (*shinkô-gangu*). Hawks are more than merely decorative, they also represent commercial success and are thought to bring luck. People like to set them up as a kind of talisman, and so they are popular presents at celebrations for new buildings, weddings and business openings.

The shavings are bent slightly back.

Hawk – knots, as seen
on this piece of wood,
are worked in.

The other birds from the workshop also have a deeper meaning alongside their decorative character. Cock and hen (*niwatori*) are early risers, standing for hard work and health, pipits and wagtails (*sekirei*) for fecundity and owls (*fukurô*) for luck.

Development into a small local branch of commerce has meant that new forms have kept emerging. Most craftsmen specialize in a small number of forms and also exchange their wares with each other.

Shavings make
a peacock's tail
(maru-kujaku).

Wooden toys from the Yonezawa region

*Touching off a
wooden doll with a
curved chisel.*

Kokeshi

One of the particular attractions of toy manufacture is its sustainability, as only young wood is used, of the kind that occurs in large quantities when clearing the woods, and for which there is hardly any other use. So it is one of the few professions working with timber where one hears no complaints about it becoming more and more difficult to source material. Similarly to Europe, most traditional wooden toys end up as folk art on shelves today, rather than being played with by children. This is most conspicuous in the case of the Kokeshi dolls, for which a real collectors' market has developed.

Kokeshi are simple wooden dolls that emerged in the latest around the mid 19th century, and spread rapidly in the north of the main island of Honshû. The dolls have a high, usually cylindrical torso, and a wide, almost spherical head, and thus a very plain, reduced form. Their charm derives from painting: the torso is decorated with rings or floral motifs, and the stylized face is highly expressive. They were originally children's toys, but soon became a popular small gift, and acquired an enthusiastic following of collectors. Today they are made in northern Japan at about 70 locations, in numerous local variants.[4]

The Niiyama workshop

The Niiyama workshop has specialized completely in making Kokeshi. The Niiyamas, father and son, work in Shirabu, a little place in the mountains south of Yonezawa famous for its hot springs. Spa tourism started a good 100 years ago, promising sufficient sales for wooden dolls. In 1926 the grandfather moved into the tiny mountain settlement, which now has only about 50 inhabitants. Collecting dolls has meant that there are real genealogies for the profession. Distinctions are made between eleven traditional schools. The Niiyamas are part of the Yajiro branch as the family comes from Yajiro near Shiroishi, a town in the Miyagi Prefecture famous for its Kokeshi dolls. The 50-year-old son Junichi, whose 78-year-old father still works a full day alongside him, is the sixth generation to practise the craft.

Working the blanks with an adze.

Turning starts – the chisel is on an oblique rest.

Knocking the blank on to the chuck.

Painting circle motifs at the lathe.

Sourcing and preparing materials

The preferred wood is *mizuki*, an ivory-white variety of maple, and also up to 20% of wild cherry (*yama-zakura*). Until 1963 the wood was obtained from charcoal-burners who lived in the villages below, and set up huts in the woods. Since the National Park was set up in 1970 it comes from more distant regions via a timber merchant from the neighbouring Fukushima Prefecture. It is felled in winter, a lorry brings 10 *koku* once a year in May, that is about 3 m³. The young timber is 3 *sun* to 1 *shaku* in diameter (9-30 cm), and is cut to length of 6 shaku (1.8 m). Niiyama still uses the old units of measurement, which were actually replaced by metric units when the country was modernized in the late 19th century, but have persisted also in this craft to the present day.

The craftsmen first remove the bark in strips, though only partially, to hurry up the drying process a little, but not too much. The logs are piled up crosswise for six months from mid May onwards. This is called *igeta wo tsumu* (*igeta* is the well crib). The stack is covered at the top, and the wind can blow through the tower, which is hollow inside. The stack is taken down in November, and the trunks are placed in a shed with their bottoms upwards, so that they dry more quickly. Mizuki is dried for three to six months according to cross-section, wild cherry for four to five years. Finally the poles can be cut to the required length with a circular saw. Then a cylinder is turned on a lathe (*ara-kezuri*). Until about 1960 the blanks were still made with a small adze (*chôna*), with the craftsman sitting at a low chopping block. If the wood is to be made into tea caddies or tops, both subsidiary products of the workshop, it is dried again for a few months on a shelf under the ceiling, to avoid later warping.

Like all Japanese turners the Niiyamas make their own turning chisels; they have a small forge close by the wooden hut for this purpose.

Working steps

The actual turning is done seated at a motor-driven wood-turning lathe (*rokuro*), whose shaft is directly in front of the craftsman. Two foot-pedals make it possible to work either clockwise or anti-clockwise. The revs can be increased continuously to 1000 revolutions. (Until 1958 the Niiyama workshop was still using a rocker-lathe (*ashibumi-rokuro*) that worked rather like a sewing machine, and was the usual turning machine in Japan from about 1890 until the post-war years.)

A circle is marked at the top of the blank with a pair of dividers. The blank is placed on the slightly conical circular chuck (*tsume*) with a few blows of the hammer; the chuck is available in four different diameters according to the size of the work-piece (10, 15, 25 and 30 mm). The turner revolves the chuck gently by hand several times to check that the blank is running concentrically. The turning chisel (*kanna-bô*) is now placed on a tool-rest called *uma* or horse with two splayed feet and slanted diagonally, and the rough shape is turned with the lathe running clockwise. Then the work is touched off with a chisel that gets slightly broader towards the blade *(baito)*. Here the left hand is used as a rest, rather than the horse. The doll is then smoothed in three steps, using grades 120 or 150, 180 and 280. When smoothing, for which horsetail (*tokusa*) was used until 1960, the turner works first anti-clockwise and then clockwise with each grade of paper, in order to create a more perfect surface.

The turner takes the doll off the shaft to paint the face.

The face is drawn with Indian ink.

The last fine lines.

The dolls' torsos are often decorated with coloured rings; these are also created at the lathe with the help of a rest. This decoration is called *rokuro-sen-moyô* in Japanese, literally lathe line pattern. The turner takes the doll off the lathe to work on the face. Eyes, eyebrows and nose are drawn with the finest Indian ink brushes, characteristically with the strands of hair on both sides. The painting is mainly in shades of red, yellow and green. It can include stylized plum and peony flowers as well as rings, and edible paints are used. The decoration rarely covers the whole doll, often large areas of the shoulder and head are not coloured. The wood looks as white as ivory here, turning cream with time. Once the paint has dried, the figure is carefully replaced on the chuck (*rokuro ni tsukeru*) and waxed anti-clockwise with Karnuba.

In the case of the little doll described here, torso and head are made from a single piece of wood (*tsukuri-tsuke*) but for larger dolls in particular a head that has been turned separately is fixed to the torso with a dowel (*oshi-komi*). Dolls whose heads can be twisted, making a squeaking noise, are called *hame-komi*. Originally the dolls were all 12 to 18 cm high, an appropriate size for children. A consequence of the dolls' status as a collector's item since the early 20th century is that they are also made as miniatures scarcely 1 cm high or as giants of over a metre.

Father and son describe themselves as *kijishi; kiji* – from the two characters for wood and ground – is a collective term for wooden cores that are further worked on by colouring or in some other way. As the Niiyamas, like other toy-makers, paint their work themselves, the term is somewhat misleading here. Father Niiyama makes 30 types of Kokeshi dolls, the son makes about 20, differing in shape, size and painting. A few tops and tea caddies, together not even 20 % of the production, round off the range. They sell all their work locally, most of it in their own shop, originally to the "tôji-kyaku", the name for the guests who used to take the waters here for several weeks. These seem to be a dying breed, having been replaced by visitors on short breaks, or day trippers, but the profession is still closely connected with tourism. The painting has not lost any of its liveliness in the Niiyama workshop, despite all the perfection of the process.

Wooden toys from the Yonezawa region

Boards for Go and Shôgi

To play Go, two players sit opposite each other at an approximately square board, with a grid of 17 lines running longitudinally and across in the early days, and 19 since the 8th century. Alternately they place their round pieces, lenticular in cross-section (*go-ishi*) on the 361 intersections (*ro*) of these lines, called *michi* or path, in order to "gain land". One player has 181 black pieces, usually made of Nachi stone; the other player has 180 white pieces, usually made of clam shells (*hamaguri*, Meretrix lusoria).

Go, which probably originated in China or India, has been known in Japan since the 8th century at the latest, and was originally played mainly by the court and clergy.[5] In the Shôsôin, the famous 8th century repository in Nara, three Go boards, two sets of pieces and several containers for the pieces have survived. The boards in the Shôsôin, of which at least the one inlaid with ebony and ivory is from China, look like low boxes whose sides have open-work ornamental patterns. A thick board with separate feet first appears on a late 12th century scroll (Kyoto, Kôsan-ji). In the 17th century at the latest the manufacture of boards for games became an independent craft in Japan, called *banya*. Demand will have increased thanks to promotion by the first Tokugawa-Shogun, who set up his own organization, *Go-dokoro*, to produce a uniform version of the game. The great popularity of the games made it possible for the profession to split further in the 19th century; there were craftsmen who specialized entirely in the elaborate feet, or in drawing the field of play. The workshops also met the demand for Shôgi boards from the early days until the present. The playing area is considerably smaller at 333 x 303 mm, with nine times nine fields marked out with black lines. This game, with its pieces made of boxwood in six different sizes, ranked by carved or painted markings, follows different rules as well. But the Shôgi boards are the same as the Go boards in shape and construction, and the craftsman follows the same sequence of steps when making them.

Under-side of a thick Go board – the carved feet are fitted into the slits at the corners. There is a V-shaped groove cut out to create a little pyramid at the centre.

208

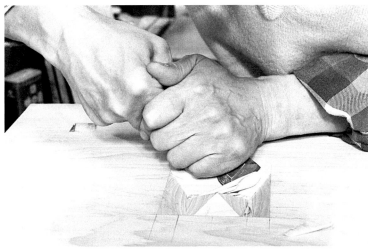

An indentation is cut out in the middle of the under-side of the block.

The Maezawa workshop

There are still about 60 little workshops in the whole country making Go and Shôgi boards. The Maezawa workshop in Tokyo is a one-man business operating in the fourth generation. It works only with indigenous timber, is one of the last to make the feet by hand and has a good reputation as the supplier to the Tenno. Until 1998 Maezawa had a workshop with retail outlet in a traditional two-storey timber building in a busy street in Hongo, in central Tokyo. But like many traditional craftsmen in large cities he was driven out of the centre as the economic pressures on property in this location increased. When the plot the family had leased for generations came to be sold, he bought a four-storey house near the Sumo stadium in the eastern outskirts of the city. He uses the ground floor as shop, store and workshop. The shop contains the large cupboards from the old business premises, and behind their sliding glass doors are board games of different qualities, waiting for purchasers. The workshop is immediately behind this sales area, and is not even 10 m² in size.

The boards can be up to 25 cm thick, and are usually made of Kaya, a conifer that can be found from southern Japan to the northern Japanese Iwate Prefecture. It provides a relatively hard wood that lasts well and turns light brown with age. Kaya wood from the south main island of Kyûshû is particularly popular. Large numbers of timber markets are held annually in the Miyazaki Prefecture, and here the workshops bid for their raw material. The timber comes from the state forest, and trunks 400 to 500 years old are needed for the best quality. Stocks have decreased considerably, so prices have risen, which is why since about 1990 timber costing only a fifth of the price has been imported from the Chinese province of Unnan.

Working stages

The timbers are cut in the sawmill under the board game-maker's supervision. The ends of the planks are immediately covered with wax to prevent shrinkage cracks, before being dried in the shed for ten to 20 years.

The manufacturing process breaks down roughly into three stages, making the actual board (*ban-zukuri*), carving the feet (*ashi-zukuri*) and drawing the field of play (*me-mori*). Maezawa fetches a well-dried plank from the stock and first prepares what is later to be

The craftsman pushes the octagonal foot against the rest with his bare right foot.

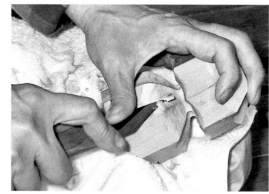

Fine work is carried out with a knife with a diagonal cutting edge; here the knife is being pushed.

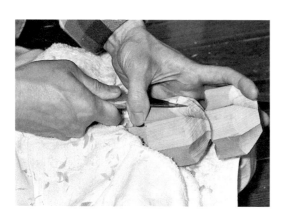

Pulling the knife; the change of direction means working with the grain of the wood.

A step is cut at the thinnest point of the foot.

When carving the foot Maezawa repeatedly plunges the short gouge into water, so that it will slide better.

the field of play and then the other sides with a roughing plane (*ara-kezuri*, coarse planing). While he is taking off thick shavings, he works the plane at an angle to the grain. As he planes, he forms a precise impression of the quality of the wood and its texture, which will be crucial to determining the price later. The heart-side is always on top of the board. If the future playing area is discoloured or has flaws in the timber he will sometimes plane the plank down further. Now the board is precisely marked out and cut. As the plank was cut somewhat larger than needed in the sawmill, the timber can be cut so that the grain on the playing area is as parallel with the edges as possible. Then all six sides of the board are planed again by hand, the top first, then the under-side, the two sides and finally the ends.

Maezawa next turns his attention to the under-side of the board. Here a V-shaped groove is cut out to create a little pyramid at the centre. This is called *chi-damari*, "here the blood collects". The background to this martial term is that no person is allowed to advise the two players during the game. But if he did, his head would be cut off and placed on the upside-down board.

Finally square slits are cut out at the four corners; these will later take the feet (*ashi-ana-bori*). Making the four feet (*ashi-zukuri*), whose shape is named after the garden fruit Kuchi-nashi, takes about two days, thus almost as long as it does to make the board. They are also made of Kaya wood, which is first planed to a square and then to an octagonal cross-section. The round plane (*soto-maru-dai*) is then used to make all eight sides slightly concave. Then a saw cut is made in the lower third (*kiri-komi*), and more work with gouges and knives follows. A thick section of beam with a recess at the

end is used as a rest. Maezawa places the blank on this, presses it against the cut-out shoulder with his bare right foot and cuts the rough shape with a gouge and hammer (*ara-uchi*). Then he follows up with another gouge and the deep indentation is carved. Finally Maezawa places a small towel over his left knee, pushes the blank against his left thigh with his right foot and holds the top end with his left hand. Work continues with a knife with an oblique cutting edge that runs to a point. This step is called *kiri-dashi*, cutting out. This knife is guided by pressing or pulling with the left thumb, according to the grain. Then a step about 1 mm high is cut at the foot indentation, this is called *heta*. The foot is then smoothed to remove traces of the knife.

The playing area, which in the case of Go boards is 18 fields (*masu*) square and for Shôgi boards nine fields square, is drawn with thin lines of black Urushi lacquer. This step, called *memori*, takes no more than an hour, but needs a great deal of concentration. So Maezawa locks himself in his workshop, which he has previously cleaned meticulously. First, markings are made in pencil (*me-wari*, dividing up). He uses two templates (*me-wari-jôgi*) for this, a rebated batten for the edging and a batten with a triangular cross-section. The latter has 19 notches on each of its diagonals, one edge for the long side (*tate*), which is 1 sun or a good 3 cm longer, the other for the shorter, broad side (*yoko*). The template is placed diagonally, so that the first and last notch corresponds with the edging line. Then the divisions are marked out on to the field of play in short strokes. Positioning the template diagonally ensures that the field of play will be marked out equally even though there might be slight dimensional tolerances in the wooden block.

Now Maezawa places the board, which can weigh up to 30 kg, on a folding wooden stand he has set diagonally in front of his workplace. He pushes some Urushi lacquer out of the tube on to a Keyaki board, adds white camphor powder (*shônô*) and mixes the two with a thin wooden spatula he has made himself. To be sure that there are no impurities in the paint it is forced through a piece of paper. To do this, Maezawa puts the paste on a piece of hand-made paper, folds this over and then twists the two ends. The paint is now literally wrung out of the strip of paper (*shibori-dashi*, wring out and flow out). The paste is then spread out on the little board over an area of about 10 x 5 cm. Oil is now applied to the edge of a wooden ruler, so that the steel spatula, a good 1 mm thick, that is used to apply the paint can slide along easily. The ruler is now placed

Marking the grid –
Maezawa moves the
spatula, whose point
he has dipped into
black lacquer, quickly
past the ruler.

in position, the spatula is pushed carefully into the prepared film of lacquer, and is then used to draw the line. The edging lines are drawn first (*me-koboshi*), then the vertical lines (*tate-sen*) in the direction of the grain and finally the transverse lines are drawn (*yoko-sen*), after the block has been turned through 90 degrees. To make it easier for players to get their bearings, a toothpick is used to mark nine positions on the field of play, four for Shôgi boards; the craftsmen call this *hoshi-uchi*, fixing on a little star.

When the paint is dry, the feet can finally be tenoned into the board. Here too the greatest precision is required, as they have to fit securely, but must not crack the board. Maezawa first fits the tenon in exactly, then transfers the outline of the foot, which is in the shape of a double quatrefoil, with the marking awl. To ensure a better fit, the marked area is chiselled about 1.5 mm deeper (*ashi-hame* or *o-ire-bori*). The completed board is protected by a fabric covering or a casing made of thin Kiri boards.

Dots called hoshi *("star") make it easier for players to get their bearings on the field of play. Maezawa applies them with a reversed toothpick.*

Prices, sales, prospects

Maezawa sells his boards in his own shop, at prices between € 1000 and € 40,000. The key criterion is the thickness and quality of the timber used. Boards with vertical annual rings and even markings command the highest prices. Slight discoloration, flaws in the timber and horizontal annual rings reduce the price considerably.

Board-making is an endangered profession. Western influences, affluence and curiosity have meant that a large number of new hobbies are pursued in post-war Japan, with a consequent diminution of the old board games' relative importance. But on the other hand, Go and Shôgi have now caught on in the West. However, changed living conditions represent the greatest threat to the profession. The boards are a good 25 cm high; they are placed on the floor, which is covered with tatami mats, and the players sit cross-legged at the board. But as most Japanese now have chairs in their homes and there is not even a single traditional room in many of them, the demand for traditional boards is decreasing markedly. So now Maezawa too makes boards that are only 4 to 6 cm thick, and are placed on a table.

Detail of the board and carved foot; boards with vertical annual rings are particularly sought after.

Drums

Japan's traditional musical instruments are classified as wind (*fuki-mono*), string (*hiki-mono*) and percussion (*uchi-mono*) instruments. While many kinds of flute and the reed organ used in court music (*gagaku*) are made of bamboo-cane, most of the other instruments are made of wood. Description of musical instrument building as a branch of the craft of woodwork will be restricted here to the three instruments that occur most frequently in traditional music: two stringed instruments, the Japanese harp (*koto*) and the three-stringed lute (*shamisen*), and from the large family of percussion instruments the large barrel drums (*ô-daiko*), which are also called *wa-daiko* or Japanese drums.[6] Japan has had barrel-shaped drums for a very long time, the oldest evidence is a 5th century *haniwa* figure from the Gunma Prefecture (*haniwa* are clay figures placed on burial mounds); it is holding a little drum under its left arm and beating it with a stick held in the right hand. There were also large barrel drums with two skins in the early Buddhist temples (e.g. in the Hôryû-ji near Nara), where they were used mainly to indicate the time. They hung in a small two-storey building (*kei-rô*), usually opposite a bell-tower (*shô-rô*) that looked almost exactly the same. But large drums in Japan are associated above all with Shinto shrines, where these drums beaten with massive wooden sticks provide an acoustic background to festivals and processions.

Making the body

The resonance chamber (*dô*) is traditionally made from a single piece of wood. The skins are made of cow- or horse-hide about 2 mm thick, which is placed on the head, stretched and fixed to the body with two or more rows of round-headed nails placed close together. There are still about 50 mostly small workshops making traditional drums in Japan. They used to be manufactured in two phases: the bodies were hol-

Aihara in his store – the edges of the drum bodies have newspaper pasted on to them.

214

lowed out in the country, delivered to town and smoothed, painted and covered there. The centre for the manufacture of drum bodies (*dô-zukuri* or *dô-bori*) was the south Aizu region (Fukushima Prefecture) from the 19th century onwards, said to have covered more than half the requirements for the whole country.[7] Until the fifties they were made in the forest, a trade called *taiko-kiji* (wooden core for drums). Just as was the case for spoons (*shakuji-kiji*) or the blanks for soup-bowls (*wan-kiji*), groups of three or four men went into the mountains, returning to their villages only at New Year, for the Feast of the Dead in August and for the harvest. Fixed workshops were set up in the 1960s, partly because of greatly reduced timber availability. They had the preferred Sen and Keyaki trunks delivered from northern Japan and Hokkaidô.

The trunks are first cut to length and marked on both ends with a simple beam compass (*bun-mawashi*). To increase the yield, especially from thick trunks, several drum

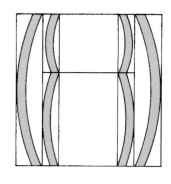

Cutting the cylinders out of the logs.

Kawada uses only 20 % of hollowed-out drum bodies now, most of his products are glued together from staves. On the left are some glued bodies, on the right, prepared staves.

215

Drums

bodies are cut from one section of a trunk if possible, and so several concentric circles are marked out. Then an 18 mm hole is drilled from both sides; great skill is needed here as the holes have to meet in the middle. Finally an extremely long padsaw (*hiki-mawashi-noko*) is introduced into the hole, and the internal radius is then carefully cut out. To do this, the craftsman sits or crouches by the log and turns it gradually. He has to guide his saw without torsion and make the cut somewhat obliquely, so that the core, which is slightly conical as a result, can be detached. In this way, up to six drum bodies can be cut from a single log, according to diameter. The outside of the cylinder is then worked on to give it its characteristic bulging shape (*katachi-zukuri*). The body is further hollowed out inside with an axe and adze, until the walls are of uniform thickness. Finally the outside is further smoothed with a curved hand plane.

An experienced maker of drum bodies can produce one drum 45 cm in diameter per day. Today only half a dozen old men in the Aizu region still practise this profession. Since the sixties they have used chainsaws to cut the trunks to size, hollow out the innermost body roughly and make a start on the outside shape. But the pad saw work on cutting into the drum body, work on the inside with a short adze and smoothing the exterior wall are all still done by hand.

Only high-quality drums are still made of hollowed-out logs (*kuri-nuki*); today most of the bodies are made by gluing staves together (*shûsei*). Drums made in this way, usually from Nara wood from Hokkaidô, cost less than half and can be made in any size, while the thickness of the trunk imposes a natural restriction on drum bodies made from a single piece of timber (max. approx. 1.2 m).

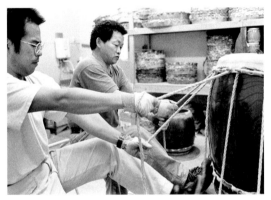

Fitting the skin

Prepared, shaped skins.

The Kawada workshop covers traditional and modern drums. Resonators for traditional drums are still obtained from the few remaining craftsmen making them, while bodies assembled from staves, which now account for 80 % of production, are made in-house. Most drum-builders buy finished skins (*ita-kawa*) from a specialists tannery in Tokyo, but Kawada tans the raw hides himself (*gen-kawa*).

Pulling on the ropes.

Before covering, the body is smoothed and the thickness of the wall is reduced to 10-15 mm inside the top. The body is painted with iron oxide stain and later with clear varnish to achieve the sought-after dark reddish-brown shade. Then two rings (*kan*) are fitted, so that the drum can be hung up.

The dry skins are first cut into circles with scissors, softened in water and then treated with the scraper on the horse (*aka-dashi*) to remove impurities from the hair follicles on the upper side (*kin-men*) and to get rid of most of the fat from the under-side (*ura* or *toko*). The skin is then scraped on the under-side with a cabinet scraper (*tachi-*

ganna) until it is uniformly thick. Then comes the temporary covering phase (*kari-gake*): The soft wet skin is placed on top of the drum body and short cuts (*me-tsuke*, adding eyes) are made at the edge with a small knife.

Four cuts in each case form the two eyes through which a steel or bamboo rod (*meku-da*) will be pushed. Ropes can now be attached to the rods and the skin stretched. Once the skin has dried it adopts the shape in a few days and is taken off again. Kawada has a large stock of skins that have been prepared in this way, piled up in his workshop like great lids.

For the actual covering (*hon-bari*) the pre-shaped skin is slightly moistened in the middle on the top. The stand used for covering (*kawa-hari-dai*) consists of a star-shaped steel frame with a number of small hydraulic jacks on it, supporting a sheet of steel. The drum body is now placed on this. Then the ropes are attached, Kawada uses old harpoon ropes from the whaling industry, as they are soft but can take very large loads. The tension can be finely adjusted by twisting the ropes using short sticks of wood or bamboo (*mojiri-bô*). The drum-builder now places wood pieces on the rods and hits with a large hammer to stretch the skin further. The tension is then reduced slightly, and immediately increased again. Finally he uses a large marking gauge to fix the position of the black-painted, round-headed nails, which are now hammered in with rhythmical precision.

If the drum is intended for a Shinto shrine or a temple, the edge of the skin with the eyes in it is later cut off with a knife. Otherwise the eyes are deliberately left, to make it easier to remove the skin and retune it finely if necessary. Kawada builds 1200 drums per year.

Jacked-up drum – the skin is fixed with round-headed nails.

Sale-room with drums of different sizes.

Covering the body.

Shamisen

Like flutes and lute, the three-stringed *shamisen* also has foreign antecedents.[8] The instrument is Chinese in origin, and became known in Japan in the second half of the 16th century via the little kingdom of Ryûkyû (Okinawa). Other instruments followed their continental models for a long time, developing in very small steps at best, but the *shamisen* very quickly started to acquire Japanese characteristics. The resonance box became bigger, and was no longer made of a hollowed-out piece of wood but assembled from four boards, and covered with cat or dog skin, rather than snakeskin. Unlike zithers, for example, which for a long time were the sole preserve of the upper classes, *shamisen* became very popular as early as the 17th century. This instrument is also unusual in having a copious and diverse literature devoted to it.

The strings are plucked with a plectrum called *bachi*, made of ivory, buffalo horn or hardwood. The instrument's particularly attractive feature is that it combines the expressive qualities of a stringed and a percussion instrument.

The instrument has a relatively small, rectangular resonance box (*dô*) and a long neck (*sao*). The body is assembled from four slightly curved mitred boards, and covered on both sides with cat or dog skin. The neck is divided into three distinct areas, the head-piece, called *tenjin* (angel) with the peg-box (*ito-gura*; string store), the long fingerboard and a slightly conical spike (*naka-gi*). The last-named part is pushed through the box like a skewer. Three strings made of twisted silk run over two bridges, of which the upper one (*kami-goma*) is mounted in front of the peg-box and the lower one (*koma*) on the resonator. The strings are secured on a string mount (*neo*) attached to the end of the conical spike at the bottom, and at the top they are wound round the tuning pegs (*ito-maki*). A distinction is made between three variants according to the thickness of the neck: a thick-necked *shamisen* (*futo-zao*) for music to accompany Japanese puppet theatre, a thin-necked version for folk music (*hoso-zao*) and a version with a medium-thick neck (*chû-zao*) for recitative songs.

The neck was originally made of a single piece of wood, but for ease of transport it started to be made in two parts from about a 100 years ago, and now usually consists of three short sections (*mitsu-ore*).

Labour is shared for the making of *shamisen*, with a specialist business at the beginning and the end of the process. This specialist business accepts the commission for the instrument, selects the wood from a timber merchant specializing in material for mu-

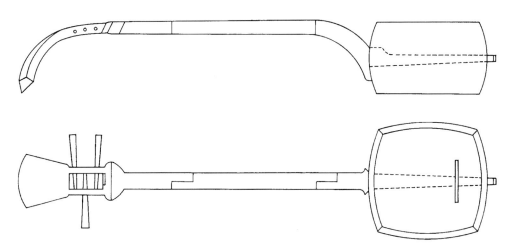

Sketch of an assembled instrument.

sical instruments, has the body and neck made by the appropriate specialists, then paints and assembles the instrument, and covers the body. Kikuoka, literally Chrysanthemum Hill, is one of about 20 little businesses in Tokyo specializing in high-quality *shamisen*. This workshop and retail outlet is a sole trader business run by Horikomi, born in 1942. The profession runs in the family, his father and his two uncles had their own workshops, and his cousins have now taken over the latter. His father's shop was not far from the Ueno station in a traditional two-storey timber building, but this was sold so that the inheritance could be divided amongst the four siblings. Since 2001 Horikomi's workshop has been on the top floor of one of the few remaining older buildings in Okachimachi, a central but less busy quarter of Tokyo. The sign on the street measures only a few centimetres, and the craftsman himself also appears unassuming. This description of resonance box and neck manufacture is based on two workshops that Horikomi co-operates with closely.

Body – dôzukuri

The resonance box is made of quince wood (*karin*, Cydonia senensis). This tree is a native of southern China, and provides a dark-red wood of medium hardness with striking markings. The four curved sides should be as even-grained as possible, with concentric rings. This essentially aesthetic ideal affects cutting and processing of the wood directly. The felled trunks are 45 to 75 cm in diameter. They are first cut into discs a good 20 cm thick; this measurement is derived from the long side of the body, whose

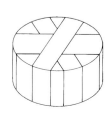

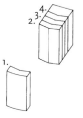

ultimate length is 195 mm. Strips about 10 cm wide are then cut out of the thick timber disc (the body is approx. 9.5 cm high), the first one from the middle, then subsequent shorter strips radially. Boards 21-24 mm thick are then cut off these blocks, which are reminiscent of a book in shape, and the boards are numbered consecutively from the outside inwards. As the sides of the body should be as uniform as possible in colour and markings, four boards from the same position are put together: four times the outside board from a block, for example. The texture becomes less clearly discernible towards the heart and the shade of colour paler, and so the outer planks are particularly sought after.

The outside curvature is checked with a gauge made of three battens.

The outside is worked with a hand plane. The wood is slightly moistened to make it easier to see where material has been taken off.

Horikomi has the bodies for his instrument made by Tsubuku. Tsubuku learned his trade in his father's workshop in Tokyo, but moved to Toda, a little village on the north-western edge of the Kantô plain 35 years ago. Here he lives in a grand but somewhat run-down large mid 19th century farmhouse that his wife inherited from her parents. The little workshop, just under ten square metres, is in front of the farmhouse. Tsubuku does not have his own stock of wood, the music shops send him four roughly cut boards from which he builds the body. Today only Karin is used; when he was younger, native mulberry (*kuwa*, Morus alba) was occasionally used because of the flourishing silk production.

The grain on the sides, with annual rings that are as concentric as possible, is crucial for the (visual) quality of the body, and its price. The craftsman will study the piece of wood very carefully to ensure that the centre of the markings is precisely in the middle, and will trim off excess material with a short-handled adze and a plane (*ara-kezuri*). He repeatedly uses templates to check the curve on the work-piece. These include very thin wooden templates that are placed on the edges, and a U-shaped gauge, made of three small, thin battens, that indicates the external curve on the sides. The boards are then planed to the required width (*haba-kezuri*); Tsubuku holds the side, which is standing on its end, with his bare feet. The outside of the body sides is further smoothed with the plane, a step called *yama-kezuri*, "planing a mountain". Tsubuku prepares the surface for this by rubbing it with a wet cloth. This makes it easier to remove the shavings and also shows him, because the moistened wood is darker in colour, where he has removed material. Next the inside is hollowed out with the small adze (*ara-chiri-tori*). Once the shape has been roughly established, the boards are marked with a thin wooden template, and cut to length on a mitre saw with the assistance of a home-made stop (*tome-setsu-dan*). They are still 2-3 mm longer than the required final size, to leave the craftsman some leeway for the rest of the work. After this the sides are hollowed out with a round-ended chisel. To do this, Tsubuku has fastened them to a heavy block of Keyaki wood, which is fixed to the floor with large iron clamps to make it more stable. The four sides are now tied together with string and hung from the ceiling to dry for one to four weeks. Tsubuku calls this *ura-boshi*, dry the back. This inter-

The inside is chiselled out.

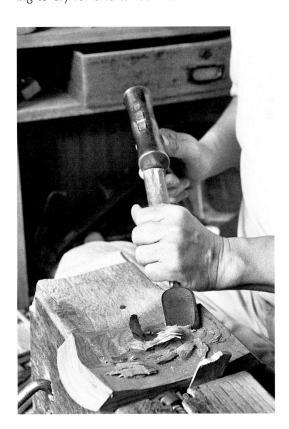

Smoothing the side mitres with a plane. The work-piece is held in place with the left foot against a cut-out piece of beam.

mediate drying phase is intended to ensure that the wood for the back, from which more material has been removed, ultimately has the same moisture content as the front.

Then the mitres on the boards are smoothed with the plane (*tome-kezuri*). The plane has a fairly broad iron, 67 mm, which is set relatively flat, to make it easier to work on the ends. The body is now provisionally assembled, and has a linen cord tied around it (*asa-nawa-kake*). Four hardwood wedges are pushed in between the cord and the body to achieve the necessary tension or pressure on the timber.

The body is dismantled again and the side strip, known as the *chiri*, about 15 mm wide, is created on the inside (*chiri-kezuri*). This is done with a knife. Then more work is done on hollowing out the inside (*ura-sarai-kezuri*), with a hammer and chisel. The work becomes gradually finer; thin shavings are taken off with a curved round blade mounted on a long handle (*nami-zuri-kezuri*), and then the concave area is smoothed with a scraper (*kosoge-kezuri*). Once a completely smooth surface has been produced the pat-

Further work with
a curved blade on
a long handle.

A pattern is cut out
on the back of the
sides; it is marked out
using sheet copper
templates.

tern to be cut out on the inside can be marked out (*ayasugi-shitaji-zukuri*). The most important measurements are transferred with templates made of thin sheet copper. The zigzag pattern, which is found on the inside of Japanese harps too, is cut with chisels with diagonal cutting edges.

Finally the mitres are planed smooth for the last time (*tome-shiage-kezuri*) and the body is assembled using bone glue. Once the glue has set and the cord has been removed, the curved outside surfaces can be smoothed as well. Tsubuku gets his younger sister to deal with the subsequent laborious polishing of the body, for which natural whetstones are used (*ara-tô*, *naka-tô* and *shiage-tô*, coarse, medium and fine). Some customers even do it themselves. Finishing with clear Urushi varnish (*urushi-kake*) is done by the specialized dealer, and the two apertures for inserting the neck (*shi-komi*) are cut there as well. Tsubuku needs three working days to complete one

Assembly of the
resonance box –
the four sides are
held together with a
rope, with wooden
wedges pushed in to
increase the pressure.

Toys and musical instruments

222

Wall with tools and templates.

body, a task he used to do in two days when younger. He is paid about 40,000 yen. Tsubuku's biggest customer is a music shop in Osaka, which takes four bodies each month. Then comes Ishibashi, a dealerin wood for musical instruments (his former employer) and two small *shamisen* shops in Tokyo, including Kikuoka.

Neck – saozukuri

Horikomi buys his necks from the Tanaka workshop in Teraomine, about 30 minutes by train west of Yokohama. Even local people have difficulty in finding their way around here; it is an area with rows of mainly two-storey buildings extending to the horizon. The workshop is unassuming and has no sign at all. It is immediately adjacent to a sixties apartment block, and with its two mats, under four square metres, it is cramped even by Japanese standards.

Inserting a silver fitting at the halving joint.

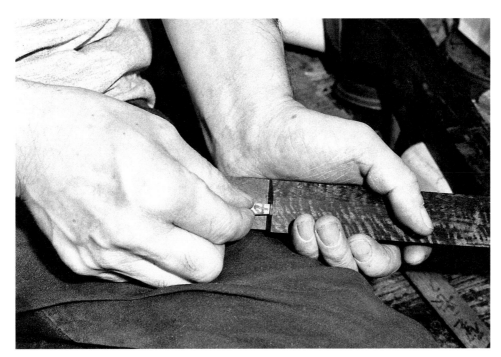

Tanaka did not take up this trade until he was 24 years old; he trained with his father-in-law for four years, setting himself up independently after he died. 90 % of his commissions come from a wholesaler in Sagamihara, the other 10 % are sporadic orders from little specialist shops like Kikuoka in Tokyo.

As with the body, the client provides the wood in this case as well. The neck and peg-box are usually made of Kôki, a dark red, heavy and very hard precious wood from the Madras region of India. Wood with an undulating grain is particularly sought after, called *tochi* or *nami-gata*.

Until the 19th century the necks were made in one piece (*naga-sao*, long neck), later in two jointed halves (*futatsu-ore*), and then from the mid 20th century they have usually been assembled from three pieces (*mitsu-ore*).

The process begins with a careful examination of the wood provided by the dealer. The luxury timber used seldom grows straight. For this reason the craftsman will lay the parts of the neck out so that he can compensate for any sections that are not straight. After marking out with the guidance of thin wooden templates he saws out the parts of the neck with a dovetail saw and a hole-saw. He then trims the halving joint known as *hada* and cuts a short square tenon at each joint. A short square metal part is placed in the corresponding mortise (*hoso-gane*). It is made of silver, or even gold for the cross-section that can be seen after fitting. Fitting this precious metal part, which is not so much for any practical reason as to enhance the value of the instrument, is not easy, as its inner edge has to lie flush with the halving joint.

A further problem when making the halving joint is that one groove or tongue has to be cut into the upper part of the joint and two into the lower one (*mizo-hori*). They provide the necessary guidance for the parts at the assembly stage, and the use of single or double groove/tongue means that the neck cannot be assembled incorrectly, and thus prevents damage. Once the joints are completed, the fingerboard is planed (*uwaba wo kezuru*).

Then Tanaka glues the spike on to the lower section of the neck, a task for which he still uses bone glue. After this the under-side of the neck is rounded off (*sao wo marumeru*). The shoulder of the neck is worked with a small knife so that it will fit precisely on to the curve of the body (*dô-shikomi*). The curve on the lower end of the neck is also created with the knife. The woodworking phase ends with making the slightly angled peg-box and the curved head beyond it (*atama wo shikomu*).

To protect the delicate longitudinal joints in the neck when it is dismantled, Tanaka makes four coverings (*kari-tsugi*) in relatively soft tulip wood. They are part of the equipment relating to the instrument, and are sold with it.

When the neck has been completed and assembled it is polished with natural whetstones (*migaki*); the stones are dipped in water, and are used in increasingly fine grades. Finally the wood is slightly moistened and repeatedly rubbed down with a cotton cloth, which gives the surface a pleasing sheen even at this stage (*tsuya-dashi*). The Urushi finish is applied by the specialized dealer, as was the case with the body.

Tanaka needs six to seven working days for a neck; polishing with the whetstones alone takes almost two days. He is paid 110,000 to 140,000 yen per neck. Until 1996 he did nothing but make necks, but the demand for high-quality instruments has fallen back considerably, and so he now delivers cakes for a confectioner in the mornings. He is to be found in his workshop in the afternoon. Commissions have increased slightly over the last three years, and his best customer is trying to persuade him to concentrate entirely on his craft again. Tanaka is not sure whether to follow his calling or the need for security.

Covering the body – the little Keyaki wood chest serves both as a work surface and tool cupboard.

The skin is stretched
with long cords
fastened to the
clamps. Hardwood
wedges are then
driven between the
two thick boards.

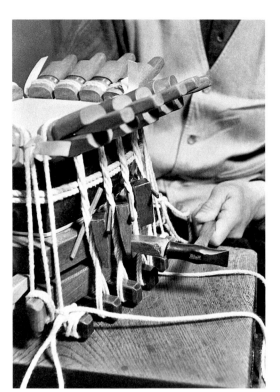

The cords can be
twisted with little
wooden or ivory rods
to increase the tension.

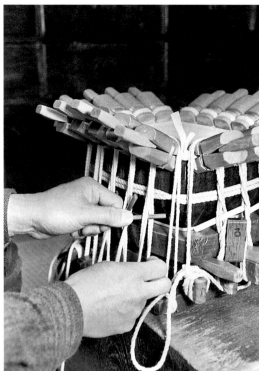

Assembly and covering

Once the body and neck have been delivered to the Kikuoka workshop, Horikomi first drills a hole in the two short sides of the body and cleans it until the spike fits precisely. He can then finish the body with transparent Urushi varnish.

The top and bottom of the body are covered with cat skin; dog skin is used for the beginners' instruments because it is somewhat thicker and more durable, and also cheaper because it can be exploited better. Ever since the Middle Ages, tanning and processing animal hides has been an occupation that has not enjoyed a high standing in this country with its Buddhist and Shintoist ideals of purity. It is still the province of a peripheral social group. They avoid publicity, all the more because animal protection organizations increasingly make their work more difficult. So despite all willingness to help, it was not possible to make contact with the last tannery in Japan. The skins for the instruments, which cost between 18,000 and 25,000 yen according to quality, have been increasingly imported from China for decades.

Horikomi's little workshop is ascetically austere and ordered. The most important tools and the stock of skins are stored in a few old cupboards with drawers. Clamps, wedges and little rods for covering are kept in the two little drawers of a cube that also serves as the work-surface.

If possible, the front (*omote*) and back (*ura*) of the resonance box are covered with parts of the same skin. Horikomi places a thin wooden template under the skin to ensure that four of the creature's eight nipples are distributed evenly across the surface. The breast skin (*kami-kawa*, upper skin) is used for the front, the rear part (*shimo-kawa*) for the bottom of the body. Then he marks the skin out with a pencil and cuts it out to a somewhat larger size with scissors. The skin is now roughened on the back (*ura wo toru*) with a narrow, slightly rounded scraper (*habuki-bôchô*) or sandpaper (grade 100), to enhance its holding qualities. The skin is then laid on a damp cotton cloth and rolled. Horikomi places the body on the moistened skin and draws round the shape with a pencil. The corners of the skin are reinforced before covering with small diagonal strips of skin called *chikara-gawa*, which are stuck on with rice glue.

The back of the body is always covered first. There are two particular reasons for this: It shows the quality of the skin, and secondly the skin that is stretched first always loses

226

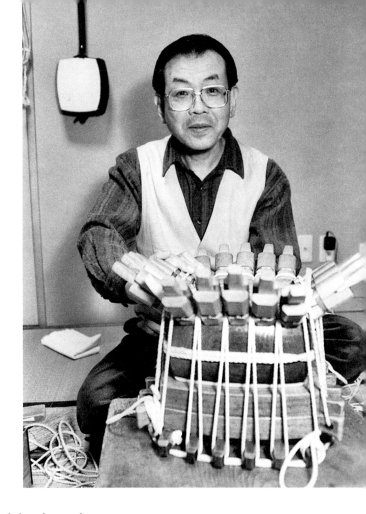

Horikomi with completed body.

Drying the freshly covered body at the charcoal brazier.

some tension. Thin strips of skin are now laid on the sides, and the clamps known as *kizen* are fitted. These are made of hard Kashi wood, and their two halves are held together with a brass wire. When little hardwood wedges have been pushed in at the back and hammered home the skin is clamped firmly at the front. First Horikomi places five of these clamps on each of the long sides, and then four on each of the narrow sides. Once all 20 clamps are in place the glue can be prepared. Rice with a particularly high adhesive strength is kneaded to a paste with a bamboo spatula on a little board. This rice is called *mochi-gome*, as it is used for the sweets called *mochi*.

Finally the rice glue is applied to the edge of the body with the spatula. The skin can now be placed on the body (*kawa wo nokeru*). Before this the craftsman places three thick wooden slabs on his work surface (*hari-dai*, a platform for covering). The two lower slabs have five and six finger-shaped extension respectively on two opposite sides. Cords are now wrapped around the clamps and these fingers on the wooden slabs (*nawa wo kakeru*), on opposite sides in each case. The tension in the cords and thus the pressure on the skin is raised in three stages. First Horikomi pulls the cords tight (*nawa wo hipparu*); to do this he has tied the end of the cord round the big toe of his right foot; he pulls on the cord tight with his right hand, catching the slack with his foot. Next, wedges are hammered in between the upper and middle wooden slabs, at the narrow ends of the body (*kusabi wo uchikomu*). These force the two slabs apart slightly, thus tightening the strings. Finally the tension can be increased again and also finely adjusted by twisting the loops in the cord with little bamboo or ivory rods. These rods, called *mojiri*, and also the hardwood wedges are stored in an orderly fashion in one of the two drawers in the work-bench. The body, covered on one side, is dried out for an hour over a charcoal brazier (*kawa wo hosu*) before the tensioning devices can be removed (*kizen wo hazusu*) and the edges of the skin trimmed precisely with a kind of marking gauge (*tachikiri*). Now the upper side can be covered in the same way.

It costs 65,000 to 80,000 yen to cover a body completely. The season is always spring and autumn, when there are a lot of concerts. A high-quality instrument costs between one and four-and-a-half million yen. Beginners' instruments can be bought for 100,000 to 800,000 yen. Horikomi sells only to end-users.

Koto – long zithers

Archaeological finds like those in Toro near Shizuoka have proved that small zithers existed in Japan as early as the 3rd century. Japanese zithers appear over the course of time in different sizes and with varying numbers of strings, from one to 17. There were direct Chinese antecedents for some variants. Until the Middle Ages *koto* was the generic term for stringed instruments, including lutes and harps. *Koto* is now understood as a long zither with thirteen strings.[9]

In the most usual version, this instrument has a resonator (*kô*) 180.5 cm long, 25 cm wide and only just under 10 cm high, made of paulownia (*kiri*). The top is curved longitudinally, laterally and also across its width. The head measures 24.5 cm, then swells to 25 cm and tapers to 23 cm at the end, known as the tail. The body is hollow inside, a board is glued in to form the flat under-side, with two ornamentally carved sound holes cut into it. Elegant feet at the corners of the body raise the instrument slightly from the floor. The curved feet at the head are a little higher than the feet at the tail, which are made of curved battens. This way the instrument is kept in a slightly tilted position. The Japanese zither has 13 strings of the same thickness; they pass over a low strip of wood set at each end of the upper side with almost the same tension. The strings are tuned with adjustable bridges (Japanese *ji*, character for *hashira* = support, stand). The bridges have little splayed feet, giving them the shape of an inverted V. The musician sits on the floor and plucks the strings with the thumb, index or middle finger of the right hand. To do this he wears plectrums (*tsume*, "claws") made of ivory or tortoise-shell, fastened to leather or paper rings.

Fujita in his wood store – the blanks are stored upright in sheds for at least six months after weathering.

Toys and musical instruments

228

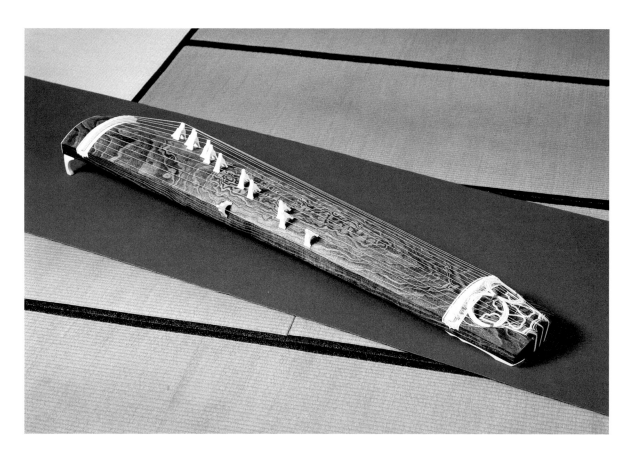

Studio photograph of the instrument, complete with strings.

Until the late 19th century local workshops were able to meet the demand for koto fully. Most of them were fortress cities, as the instrument was particularly common in Samurai families. Workshops did not concentrate in Fukuyama east of Hiroshima until the late 19th century. It is often suggested that one of the reasons for this was that two famous exponents of the instrument, Kuzuhara Kôtô and Yoshizawa Kengyô, came from Fukuyama, and also that the Kiri wood from this region was particularly sought after. The development was certainly also encouraged by the town's location, on one of the country's main transport arteries. Today about 70 % of home production comes from Fukuyama.[10] The town achieved this dominant position by sharing manufacture and early mechanization of many of the working steps. Other smaller centres of koto building are Tokyo, Niigata and Nagaoka.

Material and cutting

Kiri has been used for zithers in Japan for a long time. This deciduous tree – paulownia – flourishes all over the country, with the exception of the main northern island, Hokkaidô. The wood for the resonator should be close-grained and light; much in demand is the Kiri wood from northern Japan. However, today about 80 % of the Kiri wood used for Japanese harps is imported from Canada. The differences are subtle: Canadian paulownia has annual rings of different widths, while Japanese Kiri trees grow more evenly in terms of thickness because of the climate. They are also more resilient (*nebari*) and contain a higher proportion of oil. These qualities are also said to affect acoustics; when resonators are made of Japanese wood the sound reverberates for longer, with Canadian wood it suddenly breaks off.

The blanks for the resonator are about two metres long and a good 25 cm wide, and cost between 10,000 and 500,000 yen. The price relates above all to their grain: blanks made from relatively thin Kiri trunks are the cheapest; here the upper side of the resonator will have a grain known as bamboo shoot (*take-no-ko moku*), in other words an opposed V pattern. If the annual rings form a pattern of concentric circles, this is called *itame-moku*, and can only occur in larger trunks. The diameter of the trunk has to be even greater if the grain becomes severe and straight, *masa-moku*, as a result of vertical

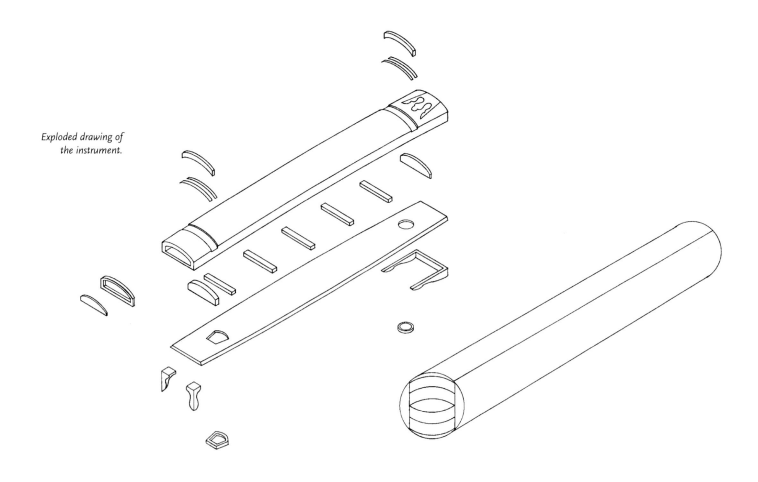

Schematic drawing of
the cutting process –
after trimming to a
width of a good
25 cm the bodies are
cut with the cylinder
mill saw.

annual rings. The resonators that are the most sought-after and the most expensive by far have little concentric circles at several points on top, because of protuberances in the trunk. Such effects are called *tama-moku*, literally bead-grain.

The blanks used to be cut until about 1965 with a two-handed saw with a specially wide blade (*maebiki-ôga*), wielded by one craftsman. A maximum of five to six resonators could be produced in one day. Today the logs are cut up mechanically. First the trunk is trimmed on two opposite sides with a band-saw to a width of a good 25 cm, then it is cut further with a so-called cylinder saw (*entô-noko*). A cylinder saw for *koto* instruments is 45 cm in diameter, the continuous cylindrical blade is 84-90 mm wide and 2.5 mm thick. The timber moves past the saw blade, which is driven by a belt on its outer side, at only 0.5 m per minute. Cylinder saws not only mean a better yield from the expensive material, they also increase production by a factor of ten, and make subsequent work on the blank easier.

The Fujita workshop

Weathering the blanks
– the exposed end has
paper glued on to it to
protect it against
shrinkage cracks.

There are seven workshops making Japanese zithers in Fukuyama, employing a total of about a 100 people. Master Fujita's one-man business is the only one in which zithers are made completely by hand, using only Japanese wood for the resonators.

The Fujita workshop is about seven kilometres from the city centre, on a hill near the port. After a six-year apprenticeship in a local craft workshop Fujita spent three years working for a harp-builder in Okayama. He then spent another year in each of two workshops in Fukuyama before establishing himself independently at the age of 26. This experience put him in full command of all the necessary working steps. This cannot be taken for granted; in the large craft workshops that emerged after the war, where instruments are made by sharing out the jobs, the craftsmen are usually involved in a particular production phase from the outset; this not only increases productivity, but sets up a hurdle for new competitors.

Fujita buys his timber, which is exclusively indigenous, costing about 40 % more than imported wood, from a small, specialized sawmill in Sôkashi in the Saitama Prefecture. He buys 80 to a 100 blanks there every year in April. He allows them to weather on the roof of his workshop for six months (*amazarashi*, literally "have them rained on"), turning them several times. This draws the sap out of the wood, above all reducing the high methanol content. This is advantageous in terms of durability, avoids later discolouring, and ensures that the wood will acquire the sought-after silky sheen (*tsuya ga deru*) as it ages. Other instrument builders, and some furniture makers, an industry where a great deal of paulownia is used, put the blanks and boards in a tank for ten days, during which the water is replaced four to five times until it no longer darkens in colour. After weathering the blanks are stored upright in shed for at least another six months.

Building the body

Building the resonator, called *kô*, for which Fujita needs about half a day, starts with the hollowing out of the under-side (*naka-guri*, "hollowing inside"). This had previously been roughly prepared with an adze, with the blank placed on the floor. Thanks to cutting with the cylinder saw, only relatively little material has to be removed today. In large workshops this too is now done with a moulder, with the exception of the final smoothing, but Fujita hollows out the shape by hand. He starts by planing a groove close to the flanks, so that he can then cut out superfluous material between the two grooves. The body is now placed on a long, narrow, slightly sloping work surface. Fujita pulls the plane

Planing the concave under-side of the body. Fujita takes out a groove on both sides before cutting out the material in between with a curved hand plane.

Fujita has placed the blank on his bench at a slant to make it easier to plane.

After the bars have been glued in a zigzag herringbone pattern is cut at both ends with obliquely sharpened chisels.

across the full length of the body in long movements. He has adapted the curved hand plane to his needs. The sole is rounded, and he has screwed a wooden grip on to the body in front of the iron, to make it easier to guide the plane. Then the sides and the top are planed; when smoothing Fujita always planes from the middle of the curved body, to prevent shaving tear-off. This laborious process is not intended only to produce an aesthetically appealing curved shape, acoustic requirements have to be met as well. Thus the wall is thicker at the head than at the tail; the body has thinner walls where the bridges for strings that are tuned particularly high are placed.

A bottom (ura-ita) 15 mm thick is fitted to the under-side of the body. It can simply be glued in – for beginners' instruments – set in a rebate, or even mitred. Once the floor has been cut to size a total of seven ribs or bars are placed on the under-side of the body. Five of these serve to reinforce the body (hara-ita, "little beam-boards"); they are set into recesses cut into the body. A "string-holder" (ito-uke) is fitted at the head. This is made of hard cherry-wood, so that the soft Kiri is not damaged by the strings. A small board (seki-ita) with several sound holes is fitted at the tail end. The string-holder and also the little board at the tail end are set in grooves in the body. All these parts have to be precisely hand-fitted because of the instrument's irregular shape.

A great deal of work goes into the inside of the body, into which a pattern intended to improve resonance is carved. These patterns used to cover the full length of the body but they are now executed only at the head and tail, covering a length of 363 and 242 mm respectively. These are precisely the areas that will later be visible through the two sound holes that have been cut in the bottom. The effort devoted to producing this primarily decorative carving is all the more surprising when one remembers that it is not possible to look into the sound holes in the floor of the instrument – at least while it is being played. The carved patterns again relate to the quality of the instrument: simple instruments have flute mouldings planed into them, called sudare-me-bori after bamboo blinds. High quality instruments are decorated with a zigzag pattern (ayasugi-bori and *komochi-ayasugi-bori*) reminiscent of herringbone. In very rare cases the so-called linen pattern is used, consisting of scored regular hexagons (*asagata-bori*). The wood is moistened before incisions 2 mm deep are made with a very sharp chisel.

Once the body has been hollowed out and the pattern cut, the bars and the bottom can be glued in. The necessary pressure is achieved by wrapping a 25 m long straw rope round the body 25-30 times. Then hardwood wedges are pushed in between the rope and the body if required, to increase the pressure.

Singeing

A special feature of *koto* building is the way the surface of the wood is treated. The resonator is singed to achieve a deep dark brown that is totally unlike the colour of untreated Kiri wood, which ages from pale white to silvery grey. This singeing or flaming process is called *yaki*. To do it, Fujita has a roofed area of about four by five metres next to his workshop. In the back corner is a low stone enclosure in which the iron tools known as *kote* are heated to about 800 °C with coke until they are red-hot, aided by a blower. The iron tools are a good 60 cm long; they are bars running to a conical tip, with a solid rectangular head. The under-side of the head can be flat (*hira-gote*) or have a slightly concave finish (*kô-gote*). The latter is needed for the curved upper side of the instrument. Once the irons are red-hot, Fujita picks them up by attaching a wooden handle a good 60 cm long. He first draws the iron briefly across a grindstone to remove the oxides that have built up on its surface while heating up. The body has already been placed on two rests. The under-side is treated first, then the long edges, for which the body is stood on its side, and finally the end sections. The body is then laid flat again, the concave iron is taken out of the fire, briefly passed across the grindstone and the singeing process takes place in four to five evenly paced passes lengthwise. The wood burns with a bright flame in the immediate vicinity of the iron. There is tension in the air, Fujita is not worried about his bare feet, on which he wears only slippers for this stage of the work as well. What is on his mind is the quality of the body, which could be rendered useless if he loses concentration. His wife helps by sweeping off any particles with a hemp palm broom. Singeing is normally carried out once a month, involving four to five resonators. They are then left in position over night.

The resonators are singed near the workshop; the fireplace for heating the irons is in the background.

233

Polishing

The next day sees the polishing phase (*migaki*) in the workshop. First, soot particles are carefully removed with a wire brush (*kasu wo toru*). Then the craftsman picks up his chalk and draws along the grain on the top, following the more porous early-growth wood precisely. The body is then rubbed down with a soft brush and the grain is again traced with chalk before the whole body is wiped with a damp cloth. After a short drying period the body is wiped one more time. Fujita now picks up a little cotton bag filled with *ibota-no-hana*. Literally translated these are privet flowers, but in fact it is a secretion that parasite larvae leave behind on privet trunks. The secretion, a kind of wax, is collected, heated and processed in fine flakes. Fujita dabs the body down with the small bag. Then it is rubbed down vigorously with a brush (*uzukuri*) made of willow roots, tied together with string to form a thin cylinder. This pushes the softer, early-growth wood in slightly, and the heat produced melts the wax; this gives the surface a relief structure, and the sought-after silky sheen. If a particularly deep glow is required, the body can be rubbed down again with a small fabric bag containing roasted rice bran (*kome-nuka*).

Decoration

The most elaborate decoration is set in the head of the body, called dragon's tongue (*ryû-zetsu*), which is slightly raised on two small, curved feet. The decoration is made of valuable wood, fishbone or ivory, usually ornamented with images in lacquer (*makie*). In Fukuyama, only the Watanabe workshop makes decorations of this kind today. All the seven remaining koto workshops buy their supplies there. The bridges are also made of valuable wood (*kôki, shitan, karin*), which the workshops order from their craft association.

The pattern of early-growth wood is traced on the singed body with chalk.

Fujita dabs the body down with a little cotton bag he has filled with ibota-no-hana powder.

The surface is vigorously treated under great pressure with a willow-root brush.

Sticking the ornamentally curved "oak leaf" on to the tail of the instrument. The necessary pressure is created by wrapping a thick string round it.

*The Fujita family –
the only daughter
works as a koto
teacher.*

Marketing

Fujita builds his Japanese zithers in small series of four to five instruments, which takes him just under a month. His good reputation means that he now sells almost all of them directly to practising musicians. He says this is no longer a financial necessity, but he wants to practise his profession for as long as possible. An instrument from Fujita's workshop costs 300,000 yen on average, but he says that occasionally he has made harps for five or six million yen. It is the decorative parts that push the price up so high; if he has feet and end-pieces made in ivory, these alone cost two million yen.

Koto are among the commonest instruments in Japan. The number of teachers alone is estimated at over 10,000. A good 40 musicians work as concert soloists. There is concern in this trade too about Chinese imports, which now make up a high proportion of beginners' instruments in particular. But overall the builders of traditional Japanese instruments are confident. This is because the Ministry of Education decided to make the study of traditional musical instruments (*hôgaku*) compulsory with effect from 1 April 2002, so the demand for beginners' instruments has increased greatly. The official return to the values of the native musical heritage represents a fundamental change in musical education, which had kept entirely to Western models since the Meiji restoration in 1868.

Strings with bridges.

The strings are plucked with plectrums.

Introduction

1 Tens of thousands of little inscribed tablets like this were found in excavations in the old capital, Nara (Heijôkyô), above all on the former palace site. The little wooden tablets containing information about casting the great bronze Buddha at the Tôdai-ji temple became particularly famous, see *Mokkan gakka* 2003.

2 One particularly spectacular example of the use of timber in civil engineering is the Sayama pond, which was constructed in the early 7th century and subsequently much extended. It provided water for the rice fields south of Osaka. Outlets that can be dated to the year 616 by dendrochronology have survived from the original structure; they are made of hollowed trunks that have been split in half. The dam was reinforced in the 17th century with a log grid, and various water release devices were added. The majority of these finds were put on show in a new on-site museum, cf. *Osaka-furitsu Sayama-ike hakubutsukan* 2001.

3 For boatbuilding, especially the boats called *sentan-bune* after their load weight of 1000 *koku* or about 150 t, particularly common in the Edo period, see *Ishii* 1995.

4 Timber for making agricultural implements, see *Inuma/Horio* 1976.

5 Toro is known above all because an especially large variety of timber objects was found here as early as 1943. Vessels and numerous items of equipment were found, as well as building components from warehouses and dwellings. Cf. *Shizuoka-shiritsu Toro-hakubutskan* 1990, 1991, 1996.

6 Extensive literature is available on the development of woodworking, especially *Watanabe* 2004, *Coaldrake* 1990, *Narita* 1984, *Muramatsu* 1973. The Takenaka Carpentry Tools Museum in Kobe holds what is probably the best collection of Japanese woodworking tools, and regularly publishes its own research.

7 *Imperial Household Agency* offered a summary of the most important objects in the Shôsôin, with descriptions in English, in 1987. Several other publications are available devoted to individual object groups, including wooden objects, see *Nihon Keizai Shimbunsha* 1978. The National Museum in Nara holds a special annual exhibition each autumn with a selection of objects from the Shôsôin, and the catalogues usually
feature a large number of wooden objects.

8 This encyclopaedia of professions is available as a reprint, *Heibonsha* 1990.

9 This survey was published as a reprint in 1982, see *Ringyô kagaku gijutsu shinkôkai* 1982.

10 *Ki wo yomu* is also the title of a book in which one of the last kobiki, or craftsmen cutting timber by hand, describes his profession, see *Hayashi* 2001.

11 For commercially available timber, processing and use, see *Zenkoku meiboku seinen rengôkai* 1986.

12 Cf. *Zenkoku meiboku shônen rengôkai* 1986, pp. 44-46.

13 The buildings at the Kasuga shrine in Nara, founded as early as the 8th century, were renewed at regular intervals of 20 years from the 13th century, for the last time in 1863. The *shikinen-zôtai* rite continued to be carried out, but the buildings were simply repaired, and not renewed. For repairs to the buildings and making new ritual tools see *Kasuga Taisha* 1996.

14 Until state monument preservation started in 1897, traces of the ageing process were often removed from shrines and temples. The Tegai-mon Gate of the Tôdai-ji and the Main Hall of the Tôshôdai-ji, both 8th century structures in Nara, are familiar examples of cutting weathered ends off rafters and also working on surfaces.

15 For a national inventory of traditional professions see for example the report for the Kyoto Prefecture, *Kyôto-fu kyôiku-iinkai* 1994.

16 For funding criteria and a summary of recognized timber professions and production centres see *Narita Mokkô sashimono* 1995, pp. 21-24.

17 14 techniques are protected in monument preservation today. For introductions to these techniques and interviews with funded craftsmen see *Seki* 2000.

Wooden bridges

[1] Introductory literature to historical bridges in Japan, see *Murase* 1999.

[2] The local museum in Iwakuni arranged an exhibition on this subject in 1998. The catalogue contains the major surviving maps, plans and images of the bridge, cf. *Iwakini Chôkokan* 1998.

[3] A detailed record of the bridge rebuilding completed by May 1953 was published, see *Shinagawa* 1959.

[4] For wooden bridges in the Alps, particularly in Switzerland, see *Blaser* 1984.

Temple construction

[5] For an introduction to Japanese religious architecture of Shinto shrines and Buddhist temples see *Hamashima* 1995.

[6] For the restoration of historic timber buildings in Japan using twelve representative monument sites see *Henrichsen* 2003.

[7] For the history of the Shôkô-ji, general plan and dimensions of the major buildings see *Takaoka-shi kyôiku-iinkai* 1994.

[8] Even the earliest surviving temples showed first signs of a modular system, as the columns' axis dimensions usually had an even value. The axis dimension of the rafters has been used as a module since the mid 13th century. It was used for determining the column positions and also the steps on the brackets and the bracket length. The gradual development of this planning system, known as *shiwari*, can be seen particularly clearly in multi-storey pagodas, cf. *Hamashima* 1968.

Workshop at the Great Shrine of Ise

[9] An outstanding pictorial record of the architecture of the Ise shrine is available, see *Iwanami Shoten* 1995.

[10] The 61st shrine renovation was recorded in several publications, see for example *Jingu shichô* 1994.

Teahouses

[11] *Sadler* 1963 and *Ehmcke* 1991 offer introductions to the Japanese tea ceremony.

[12] For a lavishly illustrated description of well-known historical and contemporary teahouses and the architecture they have influenced see *Nakamura* 1984.

[13] The materials used for building teahouses are described in detail in *Kitao* 1967.

[14] For traditional clay rendering see *Satô* 2001.

House building

[15] *Yoshida's* work first published in 1935 offers a good survey of traditional dwellings, and so does *Engel's* publication of 1964, which includes many construction details.

Architectural models

[16] See *Nara kokuritsu* 1984, p. 15. This publication also describes other forms of very small-scale architecture, like clay models of buildings and small stone pagodas.

[17] For the development of plans in Japanese architecture see *Kokuritsu rekishi minzoku hakabutsukan* 1987 and *Hamashima* 1992.

Wooden shingle and bark roofs

[18] The fullest descriptions of shingle and bark roofs are provided by two roofers, see *Tanigami* 1980 and 1982, *Harada* 1999.

Interior and furniture

House altars

[1] *Asakura shoten* 1985 provides a tabular list of manufacturing centres and a description of several regional variants, pp. 472-483.

Shôji – sliding doors

[2] *Dentô detail kenkyûkai* offers a survey of doors in Japanese architecture, 1974, pp. 95-112.

Ishô-dansu – trousseau cupboards

[3] Small chest of drawers with four sliding boxes, see *Nihon Keizai Shimbunsha* 1978, ills. 52 and 53.

[4] The Tokyo Furniture Museum catalogue offers a good survey of the various Tansu types, see *Kagu no hakubutsukan* 1986. There are also some English publications on Tansu, as a piece of furniture that is also popular in the West, see *Heineken* 1981 and *Koizumi* 1986.

[5] The Ministry of Commerce recognized Tansu from Kamo as a traditional craft in 1976. Here too the application for recognition had to have a detailed document appended; the association of furniture workshops in Kamo was kind enough to place a copy of this documentation, which is not published elsewhere, at the author's disposal.

Receptacles and tools

Boxes

[1] For the history and typology of boxes in an international comparison see *Miyauchi* 1991.

[2] For the significance of chests in connection with the tea ceremony see *Oda* 2003.

[3] For a survey of related and often very unusual shelves and small items of furniture related to the tea ceremony see *Sen* 1992.

Coopers' goods

[4] For the history of barrels and scoops including a cultural comparison between Japan, Korea and China see *Ishimura* 1997.

[5] There has been an association of manufacturers of Japanese cedar scoops and barrels in the Akita Prefecture since 1983. Their products were recognized as a traditional craft in 1984, and a documentation on history and techniques appeared in 1989, see *Akita sugi okedaru kyôdô kumia*i 1989.

Turned wooden cores for lacquered bowls

[6] A publication on contemporary lacquerware artists offers a good introduction to traditional lacquer techniques, see *Weinmeyer* 1996.

[7] For the bowl turners' working conditions and techniques see *Oku Azu chihô rekishi minzoku shiryô-kan* 2002. This publication concentrates on turners in the Aizu region, which is also known for its lacquered goods.

Chip boxes

[8] For the history and development of vessels made of bent thin boards see *Iwai* 1994.

[9] Depictions on 12th to 14th century narrative picture scrolls prove that chip boxes were widely used and that there were no stave vessels, e.g. the 1309 *Kasuga gongen engi emaki* of 1309.

[10] Boxes and sacrificial trays for rites at Shinto shrines are among the few objects still made of thin bent wood, see the two illustrations on p. 17.

[11] In the Shôsôin in Nara several 8th and 9th century round chip boxes in split cypress wood have survived, see *Nihon Keizai Shimbunsha* 1978.

Sieves and steamers

[12] *Miwa* 1989 offers a precise history of sieve manufacture.

[13] See the Teradomari local chronicle, *Teradomari-chô* 1988, Vol. 3 p. 733–736; Vol. 4 p. 450–460.

Kaba-zaiku - vessels made of cherry bark

[14] For the history of cherry bark work, typical uses and different surfaces see *Kakunodate-chô kaba-zaiku denshôkan* 1982. This was published by an establishment for exhibition and professional training spaces for local craftsmen, that has existed since 1978 in the northern Japanese city of *Kakunodate*, the centre of the cherry bark industry.

Spoons

[15] For the formal variety and function of Japanese spoons see *Akioka* 1979.

Moulds

[16] A large Japanese confectioner's exhibition catalogue provides a summary of the history of moulds, see *Toraya* 1996.

[17] For the wide range of traditional sweetmeats in *Kyoto see Kyô-gashi dokuhon*. In: *Kurashi no sekkei*, no. 196, 1990. The use of moulds for making dried sweets is shown on p. 26.

Combs

[18] The types of wood, shapes and manufacturing steps for wooden combs are described in a 1912 publication about wood used for craft purposes, see *Ringyô kagaku gijutsu shinkôkai* 1982, pp. 842–863.

[19] For the history of the Jûsanya workshop see *Nakamura* 1986, pp. 72-83.

Geta – wooden sandals

[20] *Ushioda* 1973 offers a survey of traditional Japanese footwear, for the wooden sandals described here see pp. 51-134.

[21] The only specialist museum, *Nihon hakimono hakubutsukan*, is in the little town of Matsunaga in the Hiroshima Prefecture, the centre for traditional footwear manufacture.

Toolmaking

[22] An association promoting traditional products has issued a publication describing tools from Sanjô and the most important steps in their manufacture, see *Sanjô-shi dentô shijô seihin shinkô kyôgikai* 1996.

[23] For the history of planes see *Watanabe* 2004, pp. 302-330; for international comparisons on plane manufacture see *Greber* 1987.

[24] The National Museum of History and Folklore devoted an exhibition to snap lines and building rites. The catalogue shows a lavish selection of Japanese, Korean and Chinese snap lines, see *Kokuritsu rekishi minzoku hakubutsukan* 1996.

Toys and musical instruments

Nô masks

[1] For the history of Nô theatre, its masks and costumes see *Linden-Museum Stuttgart* 1993, *Museum Rietberg Zurich* 1993.

[2] For a detailed description of all manufacturing stages including the colour finishing mentioned only in passing here see *Hori/Masuda/Miyano* 1998.

Wooden toys

[3] Saitô 1971 offers a good survey of typical regional toys. For the carved birds from the Yonezawa region discussed here see in particular pp. 354-5.

4   As Kokeshi are very collectable, there is also extensive literature, including a guide to all the manufacturing centres, see *Tsuchihashi* 1973.

Boards for Go and Shôgi

5   For the history of Go see *Masukawa* 1987.

Drums

6   For an illustrated history of traditional Japanese musical instruments see Tokyo *Shoseki* 1992.

7   For the history of drum body production in the Aizu region see *Saitama-kenritsu minzoku-bunka centre* 1990.

Shamisen

8   See *Narabe* 2004.

Koto

9   For the history of Koto instruments see *Hiroshima-kenritsu rekishi hakubutsukan* 1994.

10   The manufacture of Koto instruments was recognized as a traditional craft in 1985. The documentation drawn up then, with detailed information about history and manufacturing steps was later published in book form, see *Fukuyama hôgakki seizô-gyô kyôdô kumiai* 1988.

Illustration credits

Bauer, Roland; Braunsbach	p. 36 (top left), 39, 48 (right), 49 (bottom left), 76 (bottom left, bottom right), 77 (4), 79 (top right), 86, 88 (2), 89 (2), 90 (4), 91 (right), 93 (top left), 94 (top), 102 (2), 103 (2), 115 (5), 116 (2), 121 (bottom right), 124 (top), 125 (6), 129 (2), 131 (left), 132 (2), 133 (4), 134 (2), 135 (2), 136, 137, 139, 140 (bottom left, bottom right), 141 (bottom left, bottom right), 142 (4), 144, 145 (2), 146 (4), 147 (2), 148, 149 (2), 150 (top left), 151 (4), 153, 154, 156 (4), 157 (top right), 158, 159, 160, 161 (bottom), 162 (bottom left, bottom right), 163 (2), 164 (2), 165, 166, 167, 168 (3), 169 (top right, bottom), 170, 171 (bottom left, bottom right), 172 (bottom), 173 (bottom right), 175, 176, 177, 178, 179 (2), 180 (top left, top right, bottom left), 181 (top left, top right), 182, 183, 184, 189, 190 (2), 191 (3), 195, 196 (4), 197 (4), 198, 200, 201, 202 (4), 203 (top right, bottom), 204, 205 (top left, bottom left, bottom right), 206 (2), 207, 214, 215 (top, bottom), 216 (2), 217 (2), 228, 232 (3), 233, 234 (2), 236, 237 (2)
Chôkokan Museum; Iwakuni	p. 34 (top)
Fischer, Angelika; Berlin	p. 11, 13 (2), 15 (2), 17 (2), 19, 21, 23, 25 (2), 27 (2), 29, 117, 127, 143 (bottom)
Fujita, Fusahiko; Fukuyama	p. 229
Furniture Museum; Tokyo	p. 18
Henrichsen, Christoph; Andernach	p. 20, 24, 32, 33 (top), 34 (bottom left, bottom right), 37 (4), 46 (3), S. 47 (top right), 48 (left), 49 (bottom right), 51 (2), 52 (top right), 54 (2), 55 (2), 56 (2), 57 (2), 58 (3), 59 (3), 60, 61, 63, 64 (redrawn according to original pencil sketch), 65 (2), 66 (top left), 67 (top right, bottom right), 68 (4), 69 (bottom right), 70, 71 (2), 72, 73 (2), 76 (top left), 80, 81, 82 (3), 83 (top left, bottom right), 84, 85 (top), 93 (top right, bottom), 94 (bottom), 95 (3), 96, 97, 101 (2), 105, 106 (4), 107, 108 (4), 109 (2), 110 (3), 111 (3), 112 (3), 113, 114, 120, 121 (top), 122 (2), 123 (3), 124 (bottom left), 126, 128, 130 (4), 131 (right), 138, 140 (top left), 141 (top right), 143 (top right), 150 (top right, bottom left), 152, 155, 157 (top left, bottom left, bottom right), 161 (top), 162 (top left, top right), 169 (top left), 171 (top right), 172 (top left), 173 (top left, top right, bottom left), 174 (4), 180 (bottom right), 181 (bottom right), 185, 186 (4), 187 (4), 188 (4), 194, 199, 203 (top left), 205 (top right), 208, 209 (2), 210 (5), 211 (2), 212 (2), 213, 215 (right), 218, 219, 220 (3), 221 (2), 222 (2), 223 (2), 224 (2), 225, 226 (2), 227 (2), 230 (2), 231 (2), 235 (4)
Homare Sekkei; Fukumitsu	p. 74, 75
Imperial Household Agency, Shôsôin Office; Nara	p. 12 (3)
Imperial Household Agency; Tokyo	p. 8
Ise Jingu; Ise	p. 50, 52 (4)
Kanade, Michiru; Sawara	p. 91 (left)
Kimura, Tsutomu; Nagaoka	p. 10 (bottom)
Kintai-kyô Construction Office; Iwakuni	p. 35 (4), 36 (3), 38
Kiya, Masanao; Fukumitsu	p. 78 (4), 79 (bottom)
Kobori Butsugu; Kyoto	p. 100 (drawing and photograph)
Kyoto Prefecture, Board of Education	p. 104
Misawa, Hiroaki; Tokyo	p. 40, 41, 42
Shôkô-ji Restoration Office; Fushimi	p. 43, 47 (top left, bottom left, bottom right)
Suntory Museum of Art; Tokyo	p. 14 (top)
Wada, Yasuhiro; Akashi	p. 83 (top right), 85 (bottom)
Yamamoto Kôgyô; Kyoto	p. 62, 63 (left), 66 (bottom left, bottom right), 67 (top left, bottom left), 69 (left, top right)

Bibliography

The English translations of the Japanese titles are not literal. The aim is rather to indicate the subject matter.

Akioka, Yoshio	*Mokkô dôgu no shitate* (Setting and Using Woodworking Tools). Tokyo, Bijutsu shuppansha, 1976 (13th ed. 1991).
Akioka, Yoshio	*Japanese Spoons and Ladles.* Tokyo/New York, Kodansha International, 1979.
Akioka, Yoshio	*Mokkô to dôgu* (Woodworking Tools). Tokyo, Sanreisha, 1981.
Akita sugi okedaru kyôdô kumiai (Ed.)	*Akita no okedaru* (Scoops and Barrels from Akita). Akita 1989.
Aoki, Kusuo	*Kintaikyô no kôzô* (Construction of the Kintai-Bridge). Iwakuni 1952.
Asakura shoten (Ed.)	*Mokuchiku kôgei no jiten* (Dictionary of Wood and Bamboo Applied Arts). Tokyo 1985.
Blaser, Werner	*Schweizer Holzbrücken* (Wooden Bridges in Switzerland). Basel, Birkhäuser, 1982.
Coaldrake, William H.	*The Way of the Carpenter. Tools and Japanese Architecture.* New York/Tokyo, Weatherhill, 1990.
Dentô no detail kenkyûkai (Ed.)	*Dentô no detail* (Traditional Architectural Details). Tokyo, Shôkokusha, 1974 (7. ed. 1992).
Ehmcke, Franziska	*Der japanische Teeweg* (The Japanese Tea Ceremony). Köln, DuMont, 1991.
Engel, Heinrich	*The Japanese House – a Tradition for Contemporary Architecture.* Rutland/Tokyo, Tuttle, 1964 (13th ed. 1988).
Fukuyama hôgakki seizô-gyô kyôdô-kumiai (Ed.)	*Koto no hibiki* (Sound of the Koto). Fukuyama 1988. [Production of Koto-Instruments in Fukuyama]
Greber, Josef M.	*Die Geschichte des Hobels* (History of Hand Planes). Hannover, Verlag Th. Schäfer, 1987 (Reprint of the 1956 edition).
Hamashima, Masaji	*Tô no hashirama sumpô to shiwari ni tsuite* (Axis Dimensions of Columns in Pagodas and the Shiwari-Planning System). In: *Nihon kenchiku gakkai ronbun hôkokusho* 172, 173 (Reports of the Architectural Institute of Japan). Tokyo 1968.
Hamashima, Masaji	*Sekkeizu ga kataru kokenchiku no sekai* (Historic Architecture Seen on Original Plans). Tokyo, Shôkokusha, 1992.
Hamashima, Masaji	*Shaji kenchiku no kanshô kiso chishiki* (Introduction to the Architecture of Shinto Shrines and Buddhist Temples). Tokyo, Shibundô, 1995 (3rd ed.).
Harada, Takashi	*Hiwada-buki to kokera-buki* (Bark and Wooden Shingle Roofing). Kyoto, Gakugei shuppansha, 1999.
Hashimoto, Tetsuo	*Rokuro* (Woodturning). Tokyo, Hôseidaigaku shuppankyoku, 1979 (6th ed. 2001).
Hayashi, Iichi	*Ki wo yomu* (Reading Wood). Tokyo, Shôgakukan, 2001. [Book on the conversion of trees by hand]
Heibonsha (Hg.)	*Jinrinkinmôzue* (Illustrated Dictionary of Occupations). Tokyo, Heibonsha, 1990 (Reprint of the 1690 edition).
Heineken, Ty and Kiyoko Heineken	*Tansu* – Traditional Japanese Cabinetry. Tokyo/New York 1981.
Henrichsen, Christoph	*Historische Holzarchitektur in Japan – Statische Ertüchtigung und Reparatur* (Historic Wooden Architecture in Japan – Reinforcement and Repair). Stuttgart, Theiss Verlag, 2003.
Hirai, Kiyoshi and Tokio Suzuki	*Nihon kenchiku no kanshô kiso chishiki* (Introduction to Japanese Architecture). Tokyo, Shibundô, 1995. [Emphasis is on residential buildings]
Hiroshima kenritsu rekishi hakubutsukan (Ed.)	*Nihon koto hajime – Fukuyama he no nagare* (Origin of the Japanese Koto-instrument – Development Leading to the Production of Koto-Instruments in the City of Fukuyama). Fukuyama 1994.
Hori, Yasuemon; Masuda Shôzô; Miyano Masaki	*Nômen – kanshô to uchikata* (Nô-Masks – Appreciation and Carving). Kyoto, Tankôsha, 1998.
Ihô bunka zaidan (Ed.)	*Nihon no hakimono* (Japanese footwear). Matsunaga 1999.
Imperial Household Agency (Ed.)	*Treasures of the Shôsô-in.* Tokyo, Asahi shimbunsha, 1987 (3 vols.).
INAX (Ed.)	*Tsugite shiguchi – Nihon kenchiku no kakusareta chie* (Wood Joints - Hidden Treasures in Japanese Architecture). Tokyo 1984.
Inuma, Jirô und Naoshi Horio	*Nôgu* (Farming Tools). Tokyo, Hôseidaigaku shuppankyoku, 1976.
Ishii, Kenji	*Wabune* (Traditional Japanese Boats). Tokyo, Hôseidaigaku shuppankyoku, 1995 (2. vols.).
Ishimura, Shinichi	*Oke – taru* (Scoops and barrels). Tokyo, Hôseidaigaku shuppankyoku, 1997 (3 vols.).
Iwai, Hiromino	*Magemono* (Thin Bent Wood Receptacles). Tokyo, Hôseidaigaku shuppankyoku, 1994.
Iwakuni Chôkokan (Ed.)	*Kintaikyô ni kansuru shiryô* (Historic Documents on the Kintai-Bridge). Iwakuni 1988.
Iwakuni Chôkokan (Ed.)	*Kintaikyô-ten zuroku* (Exhibition Catalogue of the Kintai-Bridge). Iwakuni 1998. [Contains numerous historic plans and pictures.]

Iwanami Shoten (Ed.)	*Ise-jingu* (Ise-Shrine). Tokyo 1995.
Jingu shichô (Ed.)	*Jingu–dai-rokujûikkai Jingu shikinen sengu wo hikaete* (61st Renewal of the Shrine at Ise). Ise, 1986.
Jingu shichô (Ed.)	*Dai-rokujûikkai shingu shikinen sengu* (61st Renewal of the Shrine at Ise). Ise 1994.
Kagu no hakubutsukan (Ed.)	*Wa-dansu hyakusen* (Selection of One Hundred Japanese Tansu. Catalogue from the Tokyo Furniture Museum). Tokyo 1986.
Kakunodate-chô kabazaiku denshôkan (Ed.)	*Dentô sangyô kabazaiku* (The Traditional Industry of Working Cherry Bark). Kakunodate 1982.
Kasuga Taisha (Ed.)	*Dai gujûkyûji shikinen zôtai hôkokusho kinenshi* (Report on the 59th Renewal of the Shrine). Nara 1996.
Kawazoe, Noboru	*Ki no bunka no seiritsu* (Emergence of a Culture of Wood). Tokyo, Nihon hôsô shuppan kyôkai, 1990.
Kibi ko-kenchiku shûfuku shiryô chôsa kentô iinnkai (Ed.)	*Kibi-chiiki ni okeru hiwada no chôsa* (Survey on Bark Roofing in the Kibi Region). Okayama 2001.
Kitao, Harumichi	*Chashitsu no zairyô to kôhô* (Material and Construction of Teahouses). Tokyo, Shôkokusha, 1967 (19th ed. 1994).
Kiuchi, Takeo	*Mokkô no kanshô kiso chishiki* (Introduction to Working with Wood). Tokyo, Shibundô, 1996.
Koizumi, Kazuko	*Kagu* (Furniture). Tokyo, Kondô shuppan, 1980.
Koizumi, Kazuko	*Tansu.* Tokyo, Hôseidaigaku shuppankyoku, 1982.
Koizumi, Kazuko	*Traditional Japanese Furniture* (translated by Alfred Birnbaum). Tokyo/New York, Kodansha International, 1986.
Kokuritsu rekishi minzoku hakubutsukan (Ed.)	*Kozu ni miru nihon no kenchiku* (Japanese Architecture Seen on Historic Plans). Sakura 1987.
Kokuritsu rekishi minzoku hakubutsukan (Ed.)	*Ushinawarete yuku banshô no dôgu to gishiki* (Tools and Rites of the Carpenter). Sakura 1996.
Kyoto bunka hakubutsukan (Ed.)	*Kyô no takumi-ten. Dentô kenchiku no ude to rekishi* (Exhibition Catalogue of Building Trades in Kyoto. Techniques and History of Traditional Architecture). Kyoto 2000.
Kyoto-fu kyôiku iinkai (Ed.)	*Dentô no teshigoto–Kyôtofu shoshoku kankei minzoku bunkazai chôsa hôkokusho* (Working by Hand – Report on Traditional Craft Professions in Kyoto-Prefecture). Kyoto 1994.
Linden-Museum Stuttgart (Ed.)	*Nô-Gewänder und Masken des japanischen Theaters* (Nô-Costumes and Masks of the Japanese Theater). Stuttgart 1993.
Masukawa, Kôichi	*Shôgi* (The Game of Shôgi). Tokyo, Hôseidaigaku shuppankyoku, 1977.
Masukawa, Kôichi	*Go* (The Game of Go). Tokyo, Hôseidaigaku shuppankyoku, 1987.
Miwa, Shigeo	*Furui* (Sieves). Tokyo, Hôseidaigaku shuppankyoku, 1989.
Miyauchi, Satoshi	*Hako* (Boxes). Tokyo, Hôseidaigaku shuppankyoku, 1991.
Mokkan gakkai (Hg.)	*Nihon kodai mokkan shûsei* (Collection of Mokkan Dating to Japan's Ancient History). Tokyo, Tokyo daigaku shuppankai, 2003.
Mokuzô kenchiku kenkyû foramu (Ed.)	*Mokuzô kenchiku jiten* (Dictionary of Wooden Architecture). Kyoto, Gakugei shuppansha, 1995.
Muramatsu, Teijirô	*Daiku dôgu no rekishi* (History of Carpentry Tools). Tokyo, Iwanami shoten, 1973 (10th ed. 1993).
Murase, Satami	*Nihon no ki no hashi–ishi no hashi* (Wooden and Stone Bridges in Japan). Tokyo, Sankaidô, 1999.
Museum Rietberg (Ed.)	*Nô-Masken im Museum Rietberg Zürich* (Nô-Masks from the Rietberg Museum Collection). Zürich 1993.
Nakamura, Masao	*Chashitsu no rekishi* (History of Teahouses). Kyoto, Tankôsha, 1998.
Nakamura, Masao (Ed.)	*Sukiya kenchiku shûsei* (Compilation of Sukiya-style Architecture). Tokyo, Shôgakukan, 1984 (7 vols.).
Nakamura, Yûkô	*Gendai no takumi* (Craftsmen of Today). Tokyo, Kadokawa shoten, 1986.
Nakamura, Yûzô	*Dôgu to nihonjin–mokkôgu no nisennenshi* (The Japanese and Their Tools – Two Thousand Years of Woodworking Tools). Tokyo, PHP Kenkyûsho, 1983.
Nara kokuritsu bunkazai kenkyûsho Asuka shiryôkan (Ed.)	*Kokenchiku no sekai – haniwa kara kawaratô made* (Architecture on a Small Scale – from the Haniwa-Figures Placed around Burial Grounds to Small Stupas Made from Fired Clay). Nara, 1984.
Nara National Museum (Ed.)	*Exhibition of Shôsô-in Treasures.* Nara 1997.
Narabe, Kazumi	*Hôgakki zukuri no takumitachi* (Craftsmen Making Traditional Japanese Musical Instruments). Tokyo, Yamaha Music Media, 2004.
Narita, Juichirô	*Ki no takumi–mokkô no gijutsushi* (Woodworking trades – History of Woodworking). Tokyo, Kajima shuppan, 1984.
Narita, Juichirô	*Mokkô sashimono* (Cabinetmaking). Tokyo, Rikôgakusha, 1995.
Narita, Juichirô	*Mokkô hiki-mono* (Woodturning). Tokyo, Rikôgakusha, 1996.
Narita, Juichirô	*Magemono–tagamono* (Receptacles and Tools Made from Thin Bent Wood and Staves). Tokyo, Rikôgakusha, 1996.

Narita, Juichirô	*Horimono–kurimono* (Carved and Hewn Woodwork). Tokyo, Rikôgakusha, 1996.
Negishi, Teruhiko	*Jiman dekiru chashitsu wo tsukuru tame* (Building Teahouses to be Proud of). Kyoto, Tankôsha, 1986.
Nihon Keizai Shinbunsha (Ed.)	*Shôsô-in no mokkô* (Wooden Artefacts from the Shôsô-in). Tokyo 1978.
Oda, Eiichi	*Sadôgu no hako to hakogaki* (Boxes for Tea Ceremony Utensils and Inscription on Box Lid). Kyoto, Tankôsha, 2003.
Oku Aizu chihô rekishi minzoku shiryôkan (Ed.)	*Kiji monogatari* (Production of Wooden Cores for Lacquerware in the Aizu-Tajima Region). Tajima 2001.
Osaka-furitsu Sayama-ike hakubutsukan (Ed.)	*Jôsetsu tenji annai* (Catalogue of the Permanent Exhibition in the Sayama-ike Museum). Osaka, 2001.
Ringyô kagaku gijutsu shinkôsho (Ed.)	*Mokuzai no kôgei-teki riyô* (Use of Wood in Applied Arts and Crafts). Tokyo 1982. (Reprint of the 1912 edition).
Sadler, A. L.	*Cha-no yu–The Japanese Tea Ceremony*. Kobe, Thomson, 1933. (numerous editions published since 1963 by Tuttle, Vermont/Tokyo).
Saitama kenritsu minzoku bunka centre (Ed.)	*Wadaiko* (Japanese Drums). Iwase 1990.
Saitô, Ryôsuke (Ed.)	*Kyôdo gangu jiten* (Dictionary of Regional Toys). Tokyo, Tokyôdô-shuppan, 1997. [Reprint of the 1971 edition]
Sanjô-shi dentô jijô seihin shinkô kyôgikai (Ed.)	*Sanjô kaji no ude* (Tool Making in the City of Sanjô). Sanjô 1996.
Satô Kaichirô und Satô Hiroyuki	*Tsuchikabe–sakan no shigoto to gijutsu* (Wattle-and-Daub Walls – Techniques of the Plasterer). Kyoto, Gakugeishuppansha, 2001.
Seki, Mihoko	*Kokenchiku no waza, nehori, hahori – Bunka-chô sentei hozon gijutsu hojisha 14 nin no kiroku* (Techniques in Historic Architecture – Documentation of 14 Craftsmen whose Techniques are Protected and Sponsored by the Agency for Cultural Affairs). Tokyo, Zennihon kenchikushikai, 2000.
Sen, Sôshitsu (Ed.)	*Tanamono no kokoroe to atsukai* (Information on Shelves and Their Use in the Tea Ceremony). Kyoto, Tankôsha, 1992.
Shinagawa, Tasuku	*Kintaikyô saikenki* (Report on the Reconstruction of the Kintai-Bridge). Iwakuni 1959.
Shirasu, Masako	*Ki–namae, katachi, takumi* (Trees – Names, Forms and Uses). Tokyo, Heibonsha, 2000.
Shizuoka-shiritsu Toro hakubutsukan (Ed.)	*Toro iseki dai ichiji chôsa no kiroku–Shôwa 18 nen chôsa* (Documentation on the First Archeological Survey in Toro in 1943). Toro, 1990.
Shizuoka-shiritsu Toro hakubutsukan (Ed.)	*Toro iseki shutsudo shiryô mokuroku* (Illustrated Catalogue of the Finds at Toro). Toro, 1991.
Shizuoka-shiritsu Toro hakubutsukan (Ed.)	*Ki no aru kurashi* (Living with Wood – Exhibition Catalogue). Toro 1996.
Takaoka-shi kyôiku iinkai (Ed.)	*Etchû Shôkô-ji garan* (The Temple Compound of the Shôkô-ji in Etchû Province). Takaoka 1994.
Tanaka, Nobukiyo	*Kyôgi* (Kyôgi). Tokyo, Hôseidaigaku shuppankyoku, 1980. [Kyogi are thin pieces of wood often bearing a Buddhist text]
Tanigami, Isaburô	*Hiwada-buki no gihô* (Techniques of Bark Roofing). Kyoto 1980.
Tanigami, Isaburô	*Kokera-buki no gihô* (Techniques of Shingle Roofing). Kyoto 1982.
Teradomari-chô (Ed.)	*Teradomari chôshi* (Chronicle of Teradomari). Teradomari 1989 (3 vols.).
Tokyo shoseki (Ed.)	*Nihon no gakki* (Musical Instruments of Japan). Tokyo, Tokyo shoseki, 1992.
Toraya bunko (Ed.)	*Kashigata no sekai* (The World of Moulds). Tokyo 1996.
Tsuchihashi, Keizô	*Dentô kokeshi gaido* (Guide to Traditional Kokeshi-Dolls). Tokyo, Bijutsushuppansha, 1973.
Uemura, Takeshi	*Ki to Nihonjin – ki no keifu to ikashi-kata* (The Japanese and Trees. Genealogy and Use of Tree Species). Kyoto, Gakugeishuppan, 2001.
Umeda, Fusatarô	*Mokkô no dentô gihô* (Traditional Woodworking Techniques). Tokyo, Rikôgakusha, 1994.
Ushioda, Tetsuo	*Hakimono* (Footwear). Tokyo, Hôseidaigaku shuppankyoku, 1973 (8th ed. 1997).
Watanabe, Akira	*Nihon kenchiku gijutsu-shi no kenkyû – daiku dôgu no hattatsushi* (Research on the Technical History of Japanese Architecture – Development of Carpentry Tools). Tokyo, Chûôkôran bijutsu shuppan, 2004.
Weinmayr, Elmar	*Nurimono–Japanische Lackmeister der Gegenwart* (Nurimono – Contemporary Japanese Lacquerware Masters). München, Verlag Fred Jahn, 1996.
Yoshida, Tetsuro	*Das japanische Wohnhaus* (The Japanese House). Tübingen, Wasmuth, 1954.
Yoshimi, Makoto	*Mokkôgu shiyôhô* (The Use of Woodworking Tools). Tokyo 1935. (New edition, revised by Akioka Yoshio, Sôgensha, Tokyo, 1980).
Zenkoku meiboku seinen rengôkai (Ed.)	*Meiboku shiryô shûsei* (Information on Precious Woods). Tokyo 1986 (4th ed. 1995).

ARNEL TORRES

Arnel Torres 2 0 0 6